Emma Young is an award-winning science writer and a former editor on *New Scientist*. Her articles have been published in the *Guardian* and the *Atlantic*. While researching this book she learnt how to echo-locate like a bat, have an out of body experience, and trick strangers into liking her. She lives in Sheffield with her husband and two children.

Also by Emma Young

Sane

Super Senses

The Science of Your 32 Senses and How to Use Them

EMMA YOUNG

JOHN MURRAY

First published in Great Britain in 2021 by John Murray (Publishers)
An Hachette UK company

This paperback edition published in 2022

3

Copyright © Emma Young 2021

The right of Emma Young to be identified as the Author of the Work has been
asserted by her in accordance with the Copyright, Designs and Patents Act 1988.

A CIP catalogue record for this title is available from the British Library

Paperback ISBN 978-1-473-69075-2
eBook ISBN 978-1-473-69076-9

Typeset in Bembo MT by
Palimpsest Book Production Limited, Falkirk, Stirlingshire

Printed and bound in Great Britain by Clays Ltd, Elcograf S.p.A.

John Murray policy is to use papers that are natural, renewable and
recyclable products and made from wood grown in sustainable forests.
The logging and manufacturing processes are expected to conform
to the environmental regulations of the country of origin.

John Murray (Publishers)
Carmelite House
50 Victoria Embankment
London EC4Y 0DZ

www.johnmurraypress.co.uk

To my parents, Peter and Joy
– for my love of science and of books

Contents

Introduction

If you've read a popular science book before, I imagine that what you're probably expecting right now is a striking anecdote; a potted story so intriguing and enticing that it sends you plunging down the rabbit hole of the text. If so, I'm going to disappoint you with this:

This morning, I had to drag myself out of my warm bed. Feeling a little anxious, because I had a difficult work call to make at 9am, I staggered down the stairs and into the kitchen. While switching on the kettle, I reached with my other hand for a mug from the shelf. Normally, I eat porridge for breakfast. But I was starving! It would have to be eggs on toast. First, though, coffee. I poured boiling water from the kettle into the cafetière – and, *ouch!* I really should have been more careful.

As introductory stories go, it really couldn't be more mundane. But, to borrow from Lewis Carroll, I just did six impossible things before breakfast. Impossible, that is, if you subscribe to a belief that's so embedded in our culture that it's taught to every child in primary school, including my own. I'm talking about the dogma that we have five – and *only* five – senses.

We owe this model to the Greek philosopher, Aristotle. In *De Anima* (usually translated as 'On the Soul'), which dates to some time around 335 BC, Aristotle sets out sight, hearing, smell, taste and touch as *the* senses. Aristotle was concerned with matching sense organs (such as eyes and noses and tongues) with the experiences of seeing, smelling, and so on. As far as he could tell, we had five types of such organ – though he did think the skin was the 'medium' of touch, the primary sense organ being 'something else within' – and five distinct types of sensory perception.

'One might be confident that there is not another sense beyond the five,' he wrote.

This was solid work – for someone who was working more than 2,000 years ago. Aristotle was an outstanding biologist as well as philosopher, but he was of his time. Physiology was in its infancy. The understanding of the brain was basic, to say the least. (Aristotle himself believed that the brain existed to remove heat from the blood.) Centuries of research since have shown that he had a good deal more insight into our senses than our brain. Still, there isn't a sense scientist around today who would argue that we have five senses – or even close to five.

Perhaps you're thinking: well, if we do have other senses, but they're not widely recognised, they can't be that important – so maybe this book will be like those annoying travel guides that high-light all kinds of little-known but 'must-see' sights – when in fact, they're little-known for good reason.

Let's go back to that paragraph about the entirely ordinary begin-ning to my morning. We'll take a closer look at those everyday events, with an eye to how obscure, or not, the relevant senses might be.

Six impossible things? Here's what they were:

- I felt warm. That's because I, like you, have receptors in my skin and inside my body that respond to different temperature ranges. This is called 'thermoception' and it has nothing to do with touch.
- I felt anxious. In good part, that was because my brain was processing sensory signals indicating that I was facing a threat. My ability to sense my own heart beat ('cardiac interoception') was crucial for this.
- I staggered down the stairs without having to look at my feet, and without falling. I managed that because I have a) a sense of the position of my body parts in space – a kind of limb location or 'body-mapping' sense (known properly as 'proprioception') – and b) a sense of the direction of gravity and a sense of when I'm moving horizontally (thanks to my vestibular system, deep in my inner ear).
- While switching on the kettle, I also reached for a mug on the shelf. That was thanks to limb-location.

- I was starving . . . My ability to sense that my stomach was physically 'empty' directly fed into my perceptions of hunger.
- When I spilled some boiling water from the kettle onto my hand, it hurt. That's because I have dedicated damage sensors ('nociceptors') in my skin (in fact, not just in my skin). Their signals in response to the scalding led to a perception of pain.

No one would argue that feeling pain, emotion or hunger is obscure. Neither, of course, is the ability to walk down a staircase. Yet all depend on senses that are missing from the Aristotelian framework. And though it is certainly true that startling sense discoveries have been made only in the past decade, some of these 'novel' senses are about as new to science as X-rays or pasteurisation.

Yes, I'm arguing in this book that Aristotle's model is wrong, but fundamentally the same claim could have been made 100 years ago. (In fact, it was – but, as we'll see in Chapter 6, no one outside academic circles really took it in.)

So how many senses *do* we have? And *why* do we still teach our kids that we have five? ('I have to think about my five senses,' my eight-year-old told me, when he came home with a writing project for English. 'Five, Mummy, five!')

To get at the first question – how many do we have? – it's helpful to put us in our biological place. Aristotle believed that people were special, made of different stuff to animals and plants. Now, of course, we know better. And we know that the origins of our human senses can be traced all the way back to the beginnings of life itself . . .

It's uncertain exactly where we are, or when, or what precisely is emerging from the primordial soup. But some time between about 3.7 and 4.2 billion years ago, perhaps in hydrothermal vents in the deep ocean floor or in warm, volcanic lakes, the first self-replicating entities appeared on the scene. By at least 3.5 billion years ago, single-celled microbes were making pre-history.[1]

These early microbes may have been little more than bags of replicating material. But they *were* bags. They had an inside and an outside. What really distinguished them from inanimate objects was that they could detect and respond to changes in their environment.

And it was in their fragile outer membranes – their interface with the wider world – that sensing began.[2]

Mutations that enabled these microbes to detect welcome or undesirable changes of course improved their survival chances, allowing them to prosper in their own niche, or even to move to a new one. The detection of chemical and mechanical, physical changes came first. Food, toxins and excreta from other microbes are all chemicals, so sensing these was clearly invaluable. Recognising mechanical impact with something else – registering when you're touching something, or being touched – was critical, too.

Given how fundamentally important they are, it's no surprise that these early classes of chemical- and contact-sensing have persisted through evolutionary time. Like the bacterium E.coli,[3] the pot plant on your desk,[4] or your family dog, you too, register physical contact and sense chemicals of interest. Contact, or 'pressure', is, in fact, just one of your senses of touch; as we'll discover, there's more to it than this. And when it comes to 'good' or 'bad' chemicals, you can sense these using smell receptors inside your nose and taste receptors on your tongue – though, as we'll find, in various other parts of your body, too.

Early on, when life was simple, chemical and touch senses were perfectly adequate for survival. But as organisms became more complex, the kinds of questions that they could ask of the wider world as well as of their inner, bodily environment became more complex, too. Questions such as, Where are others like me? Is there food near me? Am I touching something? were soon joined by Where is 'up'? Where is light? Where am I damaged? When should I take another breath? Am I falling? Where exactly are my limbs, relative to my torso? Are the life forms around me content, or scared? Would it really be a good idea to have sex with *him*?

As we'll discover, for all these questions, there is at least one biological route to an answer: a sense. As these new senses emerged in ancestral species, they proved so invaluable that they persisted, while also being refined and expanded, through to the modern day. Like a gelatinous jellyfish drifting through the deep ocean,[5] or a rose bush,[6] you too can sense gravity. Like a meerkat exposed in open ground in the Kalahari Desert, you can sense the sound-wave signal of a screech of alarm.

To get at what a sense really is, and so how many we might have, it's helpful to break sensing down into its stages. You – or any other species – first need a 'sensor' that is triggered by a specific change. For example, if you were to go outside on a cloudy night just before dawn, at the appearance of the very first photons of light, molecules inside some of the roughly 100 million rod cells in your retina would change shape. Your rod cells are extraordinary light sensors.[7]

Then, the detection of the change has to be capable of triggering a response. For us humans, that usually means that signals from the sensor must reach the central nervous system; in most cases, they have to get to the brain. To stay with the rod cell example, that molecule shift causes signals to speed along associated neurons through the optic nerve, straight to your brain.

This process of receiving and processing an incoming sensory signal might then lead to a conscious perception. Imagine you'd walked outside not at night but on a sunny afternoon. You might immediately become aware of a blackbird perched on a branch, or the tickle of a breeze on your arm. However, conscious awareness is not an obligatory component of sensing. It's perfectly possible to sense something – to detect an important change and even mount a response – without ever becoming conscious of it. In fact, as we'll discover, some of the most fascinating and mind-altering effects of our senses occur either below our conscious radar, or as faint, background murmurings, hard to home in on, and easy to miss – but changing your world all the same.

For Aristotle, the conscious feeling state associated with a sense was important. Sight, hearing, smell, taste and touch all involve starkly different types of conscious perceptions. This is one reason his model has stayed strong for so long. Sure, a four-year-old might say, I know exactly what it's like to *see* my brother squirm when I poke him, and of course that's different to *hearing* him squeal. And something else about Aristotle's five senses – their associated organs literally stick out. That same four-year-old can readily link vision to her eyes, hearing to her ears, and so on. This also makes it easy to teach that we have five senses – without making it correct.

Another reason that Aristotle's model has persisted, despite all the evidence against it, is that, in the West, we have neglected to consider how other cultures view human sensing. For Anlo-Ewe people in southeastern Ghana, for example, the notions of *azɔlizɔzɔ* (kinaesthesia, a movement sense, which is based on limb location sensing) and *agbagbaɖoɖo* (a vestibular sense, to do with balance) are as much part of their everyday understanding of the senses as sight or hearing.[8]

If Aristotle's model is so obviously wrong, you might be wondering why scientists don't tell us how many senses we *do* have. The reason is a dull one, really: philosophers and scientists are still arguing about how to individuate 'a' sense. It is true, unfortunately, that there is no incontestable, logically rational way of delineating our individual senses. This has made it very difficult for a new model to kick out the old one and take its place. But that does not excuse the continued propagation of what we now know to be a misrepresentation. Actually, 'misrepresentation' is putting it far too mildly. It is flat-Earth wrong.

It is high time for the academic disputes to be set aside, in favour of a much more scientifically justifiable view of how many senses we truly have. And there are many reasons why we all need this – and why it has to happen now.

To know what it is to be human, we simply have to know what we can sense. We are rightly proud of our thinking prowess. But the core function of even our impressive human brain is to receive, integrate and interpret, and then react to, sensory information.[9] In fact, although conscious perceptions are not an inevitable result of all this, there is a compelling argument that consciousness evolved *because* it can benefit this process.

If we don't understand our senses, we don't understand the fundamental way in which we respond to our outer, and our inner, worlds. Sensing evolved a very long time before thinking came on the scene. We still sense first, and think later. And this explains a lot about our preferences – even why sensory metaphors are so compelling. Yes, you could describe an acquaintance as being irascible and unwelcoming, but how much more direct and immediately meaningful it would be to call him 'spiky'. Likewise, you could declare that a friend's considerate message really meant a lot to you, but that's just not as affecting as saying that you felt 'touched'.

The fact is, our many senses enable so much of our mental and physical experience. Yes, they allow us to get out of bed and descend a staircase. They also enable us to identify friends and avoid danger. To eat what we need and not what we don't. To grab a book – or an opportunity. To navigate a city. To experience horror and love. To feel that we are located *inside* a body. Even to feel that we have a distinct 'self' at all.

This book will take you on a journey through all of our senses, and the surprising things that they can do for us. It will also become apparent that none of us senses the world in the same way – and these differences can shape our preferences and personalities, relationships, health and careers.

In some cases, the differences are extreme, with impacts to match. Imagine being utterly unaware of your inner bodily state, and unable to feel love or joy. Or being able to sniff Parkinson's disease in someone else, before symptoms even appear. Imagine being able to spin around for hours on end, without feeling dizzy. Or feeling someone else's pain so acutely, it's agonising. Picture yourself so in tune with your body that you could dance the lead role in a ballet – without being able to see. Or watch a friend play 'Yesterday' on a guitar, and, though it's the first time you've heard it, immediately play it back.

This is reality for some people. But even for the rest of us, it's clear that our senses don't just inform us, they *form* us. As a science journalist who has spent twenty-five years writing often about psychology, I've come back to the senses again and again. It is new research revealing how our behaviour, our relationships, our thoughts and beliefs are affected and even directed by our sensory experiences that I find completely absorbing.

Aristotle couldn't help but tell a sensory tale fit for his time. The true story is much bigger and bolder, with jaw-dropping twists and surreal surprises. It's also one that has to be told *now* in part because our senses are under threat.

Discovering our remarkable sensory repertoire is almost like leaping from a rock pool to a coral reef – only to find that it's already being bleached. Most of us live in a world that is radically different from the one in which our senses evolved. Modern life is throwing up unprecedented challenges. Your abilities to see, hear, smell are being affected. But 'new' senses that we rely on every day are also suffering; before they can even crawl into the limelight, they are fading – with potentially devastating implications for our physical and mental health.

The good news is that there is also an abundance of evidence that we can not only protect our senses to some extent, but also train them to be better. It is possible to enhance a sense without knowing that it exists; babies and young children do it all the time. For you as an adult, though, it's undeniably helpful to be aware not only of what you have to work with, but just how flexible these senses are. To an extent, you are in control of your own sensory destiny – and, wherever possible in this book, I will describe how to take that control, and so influence practically every aspect of your life. We'll discover how learning to tune in to and enhance your many senses can improve your sex life and your sporting prowess, your decision-making and your emotional wellbeing, your eating habits and your relationships (and yes, the list goes on . . .)

The first step is to know what you've got. So below, I've made a list. Some items may not look like much. But, as I hope I've hinted strongly enough already, when it comes to our senses, appearances can be very deceptive indeed.

The human senses

In popular culture, someone with a 'sixth sense' sees dead people, or has some other supernatural perception of the world. To get beyond the documented human senses, it would be more accurate, if less alliteratively snappy, to talk about a 'spooky' thirty-third sense . . .

Sight

1. Sight, thanks to rod cells and also cone cells, which allow perceptions of colour.
2. Sensing light to ascertain the time of day. If all your rod and cone cells were suddenly removed, you'd still detect light, because of this independent sensory system, but you wouldn't see anything. Still, this is the obvious place to put this one.

Hearing

3. Hearing, thanks to the detection of 'sound' waves by the cochlea, in the inner ear.

Smell

4. Smell, for which we have a vast array of different receptors, which together form a single system for sensing 'smelly' chemicals. (I know this sounds like a circular argument; all should become clear in Chapter 3.)

Taste

Because we have five different kinds of receptor for detecting five fundamentally different groups of chemicals that affect our ability to survive and thrive, and because these receptors aren't present only in the mouth and aren't just for sensing food and drink, Aristotle's 'taste' is better thought of as five related senses. For now, it's easiest to distinguish them by the taste perceptions that they typically generate. These are:

5. Salt
6. Sweet
7. Bitter
8. Sour
9. Umami (savoury)

Touch

Touch is our 'contact' sense – but it's really a group of three senses, each of which has its own sensors and involves different responses. These are:

10. Pressure
11. Vibration
12. Gentle, slow-moving contact (the sort you get from another person)

9

Itch (pruriception)

13. Itch. It's not touch, and it isn't pain . . . it's itch. Or pruriception, as it's properly called.

Pain (nociception)

We tend to think of pain as being 'a sense', but we are capable of registering three distinct types of physical damage or potential damage, each of which can generate distinct pain perceptions. They are:

14. Dangerous temperatures
15. Dangerous chemicals
16. Mechanical damage (pinching, tearing, cutting, slicing)

However, as Chapter 11 explains, there's also a lot more to pain than these three.

Temperature (thermoception)

17. Coolness
18. Warmth

Why not just have a single 'temperature sense'? Partly because we have distinct 'warm' and 'cool' receptors, but also because their signals result in different responses. These can be physical (if you're feeling too warm you might take off a jumper, for example) or psychological (more on this in Chapter 10).

Body mapping (proprioception)

19. Body mapping, for which we have three classes of receptors, is our essential, intuitive sense of limb location, or where our various body parts are located in space. Descending a staircase, drinking from a champagne glass, playing tennis, walking blindfold across a tightrope . . . attempting any of these without this sense could be lethal.

Senses for orientation, navigation and balance
(the vestibular senses)

20. Head rotation in three dimensions
21. Vertical motion (as in a lift) – and gravity
22. Horizontal motion (as in a car)

If these three sound a little dull, that's only because they are desperately under-appreciated. Mess with them and you'll not only risk walking in deadly circles, à la the desperate teens in *The Blair Witch Project*, but you might find yourself undergoing an out-of-body experience. Whirling dervishes don't target their vestibular systems for nothing (Chapter 7).

Inner-sensing (interoception)

Some of these senses are vital not just for our ability to stay alive, but to feel emotions (as Chapter 14 reveals).

23. Heart beat
24. Blood pressure
25. Blood carbon dioxide
26. Blood oxygen
27. Lung stretch
28. Cerebrospinal fluid pH

Gut feelings: hunger and thirst – and waste

29. Plasma osmotic pressure (an indicator of how much precious water is in your body)
30. Stomach fullness
31. Bladder fullness
32. Rectal fullness

Thirty-two senses . . . It's a big leap from five. But each of these senses has its own critical impact on how we live, and – as I hope you'll come to agree – its own extraordinary story.

Part One

Aristotle's Five Senses

I

Sight

Our most dominant – but fallible – sense

> All men by nature desire to know. An indication of this is the
> delight we take in our senses; for even apart from their useful-
> ness they are loved for themselves; and above all others the
> sense of sight.
>
> Aristotle, *Metaphysics*, Book 1

For humans, as for other primates, sight has long been regarded as
our dominant sense. It gives us, at a glance, an instant understanding
of where we are, and what's coming at us – good or bad. In a way,
it acts as a kind of 'long arm', allowing us to probe our environment,
but from a safe distance.

Light detection, the basis of sight, is an ancient sense, possessed
by most living organisms. An oak tree in the local park has it. Simple
photosynthesizing bacteria in a pond have it. Since their evolution,
perhaps 3.5 billion years ago, cyanobacteria (which you may know
as blue-green algae) have been using light to make energy.

One way that modern cyanobacteria locate the light that they need
only became clear in 2016, with a chance discovery involving a genus
of the bacteria known as synechocystis. When Conrad Mullineaux at
Queen Mary University of London and his team shone a light at a
group of synechocystis under a microscope, they noticed focused
bright spots on the opposite sides of the cell membranes. Further
experiments confirmed that the entire cell functions a bit like an
eyeball. And once synechocystis has worked out where local light is
coming from, it can head in that direction by moving tiny touch-
sensitive hairs on the outside of its cell membrane.[1]

For ancient as for modern cyanobacteria, light detection was about securing energy for survival. As a sense, it has proven so beneficial that about 96 per cent of animal species have some form of it. The earliest known fossils of an actual eye date from around 520 million years ago. And a kind of 'arms race' of visual improvements has even been suggested as the driver of the Cambrian explosion – which happened about 550 million years ago – during which all the major animal groups in existence today appeared on the scene.

Developments in the eye, the organ of vision, would have helped our water-dwelling ancestors to get better at spotting food and each other, and giving predators the slip. These improvements may even have enabled them literally to see a new future for themselves – on land.

Precisely what encouraged our vertebrate forebears to make the momentous step of climbing ashore, about 385 million years ago, is debated. But in 2017, after a detailed study of the fossil record, a team of biologists and engineers reported that there was a massive increase in vertebrate visual capabilities just prior to this transition. Shortly before the move to land, eyes nearly tripled in size and they moved from the side of the head to the top. This would have made it much easier, in theory, to peek up above the surface of the water – and to look out on a whole new world. Perhaps, the team suggests, it was seeing an unexploited cornucopia of food on land – millipedes, centipedes, spiders and more – that drove evolution to come up with limbs from fins.[2]

Millions of years later, relatively slight variations in the evolution of vision may help to explain another monumental step: how our species came to be the last hominin standing. The last common ancestor of *Homo sapiens* and our sister species, the Neanderthals, lived about 500,000 years ago. Exactly what happened after that is unclear, as a flurry of recent fossil finds has complicated what had seemed to be a neat evolutionary picture. But around 430,000 years ago, Neanderthals evolved in Europe, and some migrated to parts of Asia. About 300,000 years ago, *H. sapiens* made its appearance in Africa. By about 50,000 to 60,000 years ago, groups of *sapiens* were inter-breeding with Neanderthals in the Middle East. Around 45,000 years ago, bands of these modern humans arrived in Europe, where DNA analysis shows they were also not averse to at least some

Neanderthal sex. However, just 5,000 years later, Neanderthals as a species were extinct.

Skull examinations show that Neanderthals' brains were about the same size as ours. They had tall and stocky physiques. And they had bigger eye sockets, which presumably meant bigger eyes. So why were they the ones to die out, rather than our less physically and visually well-endowed ancestors?

Neanderthals probably had bigger eyes because they evolved at higher latitudes, where light levels are low. To be able to see well, especially at dawn and dusk, they needed larger eyes than our African *sapiens* ancestors did. Their bulkier bodies were also probably an adaptation – in this case, against the cold. But again, what at first glance might appear to be only a boon had hidden costs. The Oxford team suggests that more of the Neanderthals' brain capacity than ours had to be given over to vision and body control. This, they argue, meant relatively less brain space for cognitive functions – for reasoning and thought, for establishing and maintaining complex social networks, and innovating.[3]

The vision of the immigrating *H. sapiens* may not have been as acute. But, the theory goes, this disadvantage was more than offset by their extra cognitive capacity. Ultimately, the big-eyed (and big-bodied) Neanderthals may have found it harder to cope with their harsh Eurasian environments than *H. sapiens* did, allowing us to out-compete them. Their evolutionary end was, then, in sight.

Though there are other theories as to why *sapiens* triumphed, it does seem certain that the world looked a little different to Neanderthals than it does to us. But it's also the case that the world probably looks at least a little different to you than it does to me – and perhaps even radically different.

The foundations of vision begin early in gestation.[4] For it to develop fully, a baby needs practice at seeing.* A newborn's visual acuity is only at about 5 per cent of an adult's, and it can't see much further

---

* In fact, according to recent research at the University of Durham, some elements of visual processing, such as depth perception, don't become 'adult-like' until a child is between ten and twelve years old.

than about thirty centimetres (about the distance it would be from a parent's face if being cradled in the arms). However, it can distinguish between very dark and light shades, and see intense patches of red. By about two months, a baby can distinguish a vibrant green from a vivid red. A few weeks later, it can do the same for intense blue and red.[5]

Red, green and blue are some of the first colours that a baby learns to see because along with rod cells, which allow vision in dim light, our retinas are packed with three types of cone cells. These cone cells are packed in their millions in the fovea, in the centre of the retina.

'Blue' cone cells contain a type of opsin (a light-sensitive protein) that most readily absorbs light in the blue/violet, short wavelength part of the visible spectrum. The opsin in 'green' cone cells responds most strongly to middle wavelength green light, while the opsin in 'red' cone cells is most sensitive to light in the light-green/yellow/orange part of the spectrum, though they also detect much longer wavelengths of light, which we see as red.

It's thought that our blue-sensitive opsin started out as an ultraviolet light detector, only switching roles some time early on in mammalian evolution. Still, it didn't switch roles entirely; though we can't normally see UV light, our blue opsin is still sensitive to it.[6] The cornea and lens absorb this light before it can reach the retina, but people who have had their lenses removed because of cataracts sometimes report seeing patterns on flowers, and seeing objects that previously appeared black with a violet tinge. It's even been speculated that this is why the later paintings of Claude Monet, who had surgery for cataracts on his left eye at the age of eighty-two, are full of violet and blue.

Until some time between thirty and forty-five million years ago, our ancestor species had only red and blue opsins. Then the gene for red opsin was duplicated, and mutations made it sensitive to 'green' wavelengths. What drove this? Some researchers think it helped in spotting reddish fruit against green leaves. But whatever caused it, the change was momentous, taking the number of different colours that could be distinguished from about 10,000 to more like a million. Thanks to patterns of signals from all three types of cone cell, you can see a staggering array of shades, from the palest ivory, through magenta, to jet.[7]

Since we have three types of cone cell, we humans are 'trichromatic'. Well, most of us are. But colour vision deficiency, caused by a faulty opsin gene, is common.[8] Although full-colour blindness is rare, red–green colour vision deficiency affects about one in twelve men and one in 200 women with northern European ancestry. (It's less common in most other populations that have been studied.) It means that reds and greens can be confusing. In fact, it's for this reason that Facebook's signature colour is blue. Founder Mark Zuckerberg has said that his red–green colour blindness means that blue is the colour he sees most vividly.[9]

Though no one with typical vision can know exactly how someone with no red cones sees the world, it's thought that it varies from blue to white to yellow, with no reds or greens. People without the gene for green opsin are thought to have a similar visual experience, though red objects will look brighter to them.

One of the earliest known references to any kind of colour deficiency dates from 1794, and a lecture given by the British chemist John Dalton. 'I have often seriously asked a person whether a flower was blue or pink, but was generally considered to be in jest,' Dalton told his listeners. He suspected that the vitreous humour – the fluid inside his eyeballs – might be tinted blue. With his permission, his eyes were cut open after he died. His vitreous humour was found to be clear. It was only in the 1990s that his DNA was analysed, and the green opsin gene was found to be absent.[10]

A lack of blue cones, which causes blue–yellow colour vision problems, is less common, affecting about one in 10,000 people. Their world is thought to appear in shades of red, white and green.

Though three-cone-type, trichromatic vision is standard, there are cases of women with a fourth cone type.* This doesn't necessarily always translate into different colour perceptions. But if the fourth cone cell type is significantly different in its responses to light, compared with all of the other three, it can. Gabriele Jordan at Newcastle University has identified a woman with a fourth type of

* They are always women because of the pattern of inheritance of cone cell genes.

cone cell in the long-wave yellow/orange zone of the spectrum, for example. This extra 'yellow' cone meant that, in tests, her ability to tell apart a red/green mixture from a straight orange was far superior. She could see differences in hues that are simply invisible to most people.[11]

But even to people with typical colour vision, colours do not all look the same. A US-based team has found huge variation in the gene for red opsin. When the researchers studied this gene in 236 people from around the world, they discovered a total of eighty-five variants, or versions. These variations probably affect actual red–orange perceptions, meaning that the same 'red' apple likely looks a little different to me than it does to you.[12]

Rods for vision at low light levels and cones for colours . . . Even as recently as my own student days, this was the full story of retinal-sensing. Our eyes were for seeing – and these were the sensors that allowed us to see.

Well, it turns out that this is only part of the eye's story.

No doubt, you'll have heard of your 'body clock'. In fact, you have several clocks, which help to coordinate everything from waking to digestion. But the master clock lies in the brain, in the hypothalamus, a region that is critical for our basic life functions. To operate efficiently, this clock needs to know when day is breaking and night is falling, and it gets this information from the eye – but, it turns out, not via the sensor proteins that allow us to see.

In 1998, the German-born neuroscientist Ignacio Provencio discovered 'melanopsin', a completely different light-sensitive pigment, in the skin of the African clawed frog.[13] Within two years, he'd shown that our human retina contains it, too.

Experiments have revealed that animals who lack rods and cones, and who are blind, still sense light levels using melanopsin in their retinas, and use this information to control regular daily biological rhythms. This control is known to be important not only for sleep but for physical and mental health – as, for example, research on shift workers has shown. A mutation in our own gene for this protein has even been linked to Seasonal Affective Disorder;[14] sufferers of SAD experience low mood and depression during the dark winter months.

To help your hypothalamus to understand when the day starts and ends, it's important then to expose your eyes to bright light in the morning, but not in the evening. Michael Terman, who heads the Center for Light Treatment and Biological Rhythms at Columbia University, has various tips to help this system work as well as it can. If you can, walk to work – and try to avoid wearing sunglasses. In the home, use plenty of bright lights – but make them dimmable, so that as evening approaches, their intensity can be reduced. In Terman's experience, boosting light exposure during the day can help reduce the mid-afternoon or early evening fatigue that affects so many of us, and, when followed with low levels in the evening, also make for better sleep.[15] This is true for many blind people, too; the discovery of melanopsin led to a recommendation not to wear dark glasses.

The eye, then, is not just an organ for seeing. It's also an organ for sensing one of the most important environmental changes that we, along with a vast range of other organisms, need to track to survive and thrive: the cycle of day and night.

Although the eye, as Aristotle noted, is the sense organ of sight, we don't see in our eyes, but in our brains. And some of the most striking differences in how we humans see the world are down to variations in how our brains handle visual information.

Let's take a closer look at that journey. You've just woken up and drawn the curtains, and light floods into your bedroom. As the light stimulates your rods and cones, electrical signals speed along the optic nerve to the brain. Their first stop is the thalamus, a little structure sitting just above the brainstem that acts as our sensory relay hub. One of the thalamus's main jobs is to send incoming sensory information (except for smell signals) on to the appropriate parts of the cortex, for further processing.[16]

Signals from the retina are despatched straight to 'V1' – a skinny sheet of tissue that constitutes our 'primary visual cortex'.[17] Different populations of neurons within V1 respond to particular things. Some, for example, react to edges or lines at a particular angle – the vertical line of your curtains, say, or the right-angle of your bedhead or wardrobe. From V1, visual information is also fed on to other regions of the visual cortex, for, among other things, the processing of colour, movement, shapes and faces.[18] If you now turn and see that your daughter, say, rather than your partner, is beaming at you from over the duvet, that's because your 'fusiform face area' has received already partially-processed visual information. This little area of the visual cortex handles the recognition of a 'face' – though it needn't be human; it will also react to animal faces and even cartoon faces.[19] (Some animals, such as dogs, have regions that respond to human faces, too.[20])

People whose eyes work perfectly well but who, because of genes, injury or disease, have faults in parts of their visual cortex may be blind to stationary objects, but able to sense motion; or be able to identify a nose in a photo as a nose, but unable to perceive faces. But much less drastic differences in how our individual brains process visual signals can make for relatively subtle, but no less arresting, variations in our literal and metaphorical world-view.

For some of us, the colours of objects seem consistently less bright – less 'saturated' – than for others. People with major depression fall into this category. Variations in personality, too, have been linked to discrepancies in seeing. In particular, scoring highly for openness – a personality trait that entails curiosity and open-mindedness, and also the one that's most often tied to creativity[21] – is associated with an idiosyncrasy in how the brain deals with images coming in from each eye. The psychologists who devised the popular five-factor model of personality – with agreeableness (basically how nice you are), conscientiousness, neuroticism (emotional stability is the opposite of high neuroticism) and extraversion (with introversion at the far end of this spectrum) completing the line-up – wrote that openness is associated with a tolerance of ambiguity. And this seems to hold right down to the level of sensory perception.

A study that involved what's known as binocular rivalry provides evidence for this. Imagine that a circle of horizontal red stripes is being held up to your left eye, while a circle of vertical green stripes is being shown to the other. The brain will typically alternate the perception, suppressing first one image and then the other, so that you perceive a back and forth between the two. However, occasionally there'll be a blurring of the images, creating a blend of both.

When Anna Antinori at the University of Melbourne tried this with a group of students who'd also completed personality tests, she found that those who ranked highly for openness saw a blended image much more of the time than those who'd ranked low. The team concluded that more open people actually see the world differently.[22]

This work, published in 2017, was the first to link fundamental variations in visual perception to an aspect of personality. It could represent evidence that the brains of more open people work a little differently: the neural processes behind the increased blending may also somehow relate to open people's superior performance at divergent thinking. Being good at this, and able to come up with more possible solutions to a problem, has been linked directly to creativity.

There are other disparities in how we humans see that are also common to groups of people. However, they relate not to opsin genes, personality or socioeconomic circumstances, but to something else entirely. Research shows that, remarkably, entire groups of people can see the world differently not because of their genes, but because of their culture. (These differences are not deficits, and that point is worth stressing.)

A one-time travel agent, Debi Roberson got into academia relatively late in life. Aged forty-four, she left her teenage kids at home in the UK to embark on an anthropological investigation in a remote northern region of Papua New Guinea (PNG). ('It wasn't *because* I had teenage kids at home,' she insists.) Roberson hoped that her study might form the basis for a doctorate. What she discovered would shake academic understanding about how we perceive colours to its core.

Roberson set out expecting to find data in support of the prevailing theory that people from all over the world slice up 'colour space' in fundamentally the same way. So, there are a number of different hues that I or anyone else would categorise, and see, as 'red', and others that we all would see as being different but related – and all 'green', for example.

English is considered to contain eight basic colour terms – words that everyone uses and readily understands: red, pink, brown, orange, yellow, green, blue, purple; as well as white, black and grey. Roberson wanted to study colour perceptions among people whose language had fewer. She didn't really know where to go to find them, but she got talking to a couple of actors who lived near her home in Suffolk and they gave her a hint. They told her they'd been out to northern PNG to put on mime shows to encourage local people to use mosquito nets. They hadn't heard much talk about colour.

This was enough for Roberson. In 1997, she flew first to Port Moresby, then travelled up-country, in a local mission plane and then by canoe, to villages that were home to a previously unstudied hunter-gatherer tribe, called the Berinmo. She took with her a bed roll, food supplies, emergency medicines, kerosene – and a solar-powered light box, plus 160 coloured chips.

Using these chips, Roberson explored the Berinmo's basic colour terms. She found that, in contrast to English's eight, they had five. They didn't have different category terms for blue and green, as in English. But they did fundamentally distinguish between two types of what I would call green, which they split into *nol* and *wor*. These terms correspond to the colours of fresh versus old tulip leaves, which they consider tasty, or not, respectively. 'Think about the difference between a bright vibrant green and a khaki,' Roberson says. Of course, I as an English-speaker can see a difference between a *nol* green and a *wor* green, but for the Berinmo, anything that I could simply call 'green' could only be *nol* or *wor*. There is no single term that encompasses both. The blue of a lake or a clear sky, however, falls into the *nol* category.

Roberson then ran an experiment. She showed Berinmo volunteers a colour chip, took it away, then showed them a pair of colour chips, and asked them to pick out the original. For each pair, the degree of difference in hue between them was the same. However, sometimes both the original and alternative chip belonged to *either* the *nol* or *wor* category, or they belonged to what an English speaker would call either 'blue' or 'green'. Sometimes, they fell into different colour categories in English but not Berinmo (so if the original was blue, say, the other chip in the pair was green), or in Berinmo but not English (if the presented chip was *nol*, for example, the other chip was *wor*).

Roberson found that when the colour categories were different in Berinmo, her volunteers did better than when they were different in English. They found it a lot easier to identify a *nol* chip when given a *nol* and a *wor* chip to choose between than to identify a blue chip when shown a blue and a green chip. English speakers, whom Roberson tested back at Goldsmiths College, University of London, showed the opposite pattern.

If colour universals exist, and, perceptually, we all divide up colour space in the same way, language shouldn't matter. But it did. The resulting paper, published in the journal *Nature* in 1999,[23] sent seismic waves through the research community.

Roberson and her PhD supervisor, Jules Davidoff, along with Ian Davies at the University of Surrey, then went to test colour perception more directly. This time, partly because it was logistically easier, they studied a closer, semi-nomadic society, the Himba of Namibia.

Like the Berinmo, the Himba have a single colour term that encompasses blue and green. Using computer-based tests this time, Roberson and her colleagues found that, when briefly presented with colours arranged in a circle, Himba people struggled to spot a patch that to me would obviously be 'blue' among others that were 'green'. However, they had no trouble identifying an odd one out when it fell into a different colour category in their own language.[24]

Other research teams have since gathered a wealth of evidence that supports the idea that language influences the colours that we see. Some of these studies have involved languages in which blue is fundamentally divided. In Russian or Greek, for example, an object can't be 'blue' – it *has* to fall into the category of 'light blue' or 'dark blue', for which these languages have individual words.[25]

No one is suggesting that, when they're allowed to look closely, a person's ability to distinguish between colours and shades is determined by whatever language they speak. Ancient Greek may not have had a word for 'blue', and Homer may have famously described the sea as 'wine-dark', but that certainly doesn't mean he couldn't *see* what I would call blue. (Ancient Greek-speakers just didn't see the need to describe anything blue as blue, while the *nol* and *wor* categories exist not because the Berinmo are fond of greens but because the two terms usefully distinguish a nutritive from an old food plant. In fact, many colour terms in English are thought to have originated in a similar fashion.)

Roberson herself remembers one elderly Berinmo woman who provided grumpy, if also entertaining, evidence that she could tell apart a vast spread of colours with ease. 'She went through my set of 160 individual chips and managed to think of an insult for everybody in the village to go with every single one,' Roberson recalls. 'She'd look at one and say something like, "Oh, that's a sickly colour, it looks like my daughter-in-law's skin!"'

However, in providing evidence that culture, via language, influences visual perceptions, this research has challenged long-held ideas about how we see. Other research, such as a simple geometric shape test, supports this, too. For example, Yoshiyuki Ueda at Kyoto University, Japan, and colleagues studied groups of people from Canada, the US and Japan, showing them simple geometric shapes, such as straight lines, and asking them to find the odd one out. Sometimes one line was shorter, or longer, than the others – or one was straight while the others were marginally tilted. The team found that North Americans took longer to identify the odd one out if it was shorter than the others. This was not the case for the Japanese. But the Japanese volunteers had to look harder to identify a straight line among tilted ones than the North Americans did. Why?

The team thinks it could be to do with differences in their written languages. In East Asian writing, many characters are distinguished by subtle differences in stroke length, while in Western alphabets, slight angular alterations in letters are important. It seems that experience trains our brains to handle the kinds of visual information we usually encounter. When that training is consistent within a culture, the effects can be seen at a population level.

There is still some debate about the impacts of language on what we see. Not all researchers think it's been shown incontrovertibly that language – a 'high-level' process in the brain – has a 'top-down' impact on our sensory perceptions. But the idea that it can fits perfectly well within a very persuasive model of how we experience the world. This is the 'predictive processing' theory of perception.

According to this model, what you see, hear, smell, and so on represents your brain's current 'best guess' about what's happening.

In generating that best guess, your brain uses data streaming in from sense organs, but also expectations, based on past experience. If the sensory information seems fuzzy or unreliable, your brain may prompt you to try to get better data (to turn your head a little, for instance, or to move closer to an object). If you just can't do that, your brain will give more weight to its predictions in generating a perception. In some cases, it will even fib to you, for your own good.

'The goal of our visual perception is not to give us an accurate picture of the environment around us but to give us the most useful picture,' neuroscientist Duje Tadin at the Center for Visual Science at the University of Rochester, USA, has commented. 'And the most useful and the most accurate are not always the same.'[26]

In fact, many famous optical illusions rely on the constructive way in which our brains generate the 'most useful', rather than the 'most accurate', visual picture of the world around us.

One of the best-known in relation to colour perception was devised by American neuroscientist Edward Adelson in 1985. Adelson computer-generated an image of a green cylinder at the corner of a chequerboard of light grey and dark grey squares. A 'light' square in the path of what looks to be a shadow cast by the cylinder is in fact the same shade of grey as a 'dark' square outside the shadow.[27] Your brain, accustomed to the effect of shadows on shades, makes adjustments for shadow in generating its 'best guess' (your perception) about the colours of the squares. If your brain didn't take into account light intensity in daily life, you'd quickly get confused. A bus travelling along a street would change colour every time it passed in and out of shadow. A piece of paper would seem to be a totally different colour at dusk, compared with midday. In a situation in which we all expect a shadow, our brains all tend to rush to make the same assumptions. For this reason, the chequerboard shadow illusion gets us all in the same way.

Another illusion, one of the most famous in psychology, helps to reveal just how active, rather than passive, sensory perception can be. It's known as the McGurk effect, after Harry McGurk, a Scottish psychologist. In the 1970s, McGurk and his research assistant seren-dipitously discovered that when most people see someone's lips making the speech sound 'ba' but are simultaneously played someone saying 'ga', what they hear is neither sound – but 'da'. This effect neatly demonstrates that in processing speech, we combine visual and auditory information, to construct a perception that is not a simple reflection of either.

The great American psychologist William James wrote in his 1890 textbook *The Principles of Psychology*: 'whilst part of what we perceive comes through our senses from the object before us, another part (and it may be the larger part) always comes out of our own head.' Anil Seth, professor of cognitive and computational neuroscience at the University of Sussex, sums it up: 'The world we experience comes as much, if not more, from the inside out as from the outside in.'[28]

In Seth's own research, he's found that this is true for peripheral vision. Our representations of objects in the centre of our visual field, where our gaze is focused, are usually accurate and detailed. This isn't the case for objects around the edges. In fact, much of our visual sensory evidence of the world is sparse – and yet we generally tend to feel that we can see everything around us sharply. Seth and his colleagues have found that this apparently detailed peripheral vision is partially hallucinated. We use what's in the central region of our vision (the signals we can trust) to build up our perceptions of what's around the edges (for which the data are more fuzzy).[29]

Sometimes, we can even fail to see things that are right in front of our eyes. The best-known evidence for this comes from an experiment so frequently discussed it's now often referred to by psychologists simply as the 'gorilla study'. Participants were asked to watch a video of two teams playing basketball. One team was dressed in white T-shirts, the other in black. The viewers were asked to count the number of passes between the players in white. At one point, a researcher dressed in a gorilla suit wandered through the game. Most of the viewers completely failed to notice.

The explanation? Our capacity for conscious attention is limited. There's only so much that can feed into what we consciously perceive at any one time. When we're at our limit, we can experience what's known as 'inattentional blindness', and fail to perceive even highly unexpected sensory information. In this case, the retinas of the participants in that study responded to the appearance of a person in a gorilla suit, but their brains did not deem the information important enough to make it into consciousness.

That last point is important. If you were taking part in that experiment in the flesh, not watching the players on a screen, and an *actual* gorilla wandered onto the court, I'd bet my house that your brain would alert you to it. All kinds of non-threatening stuff can happen in the background, and that's fine. But a dangerous animal – or someone's eyes on you – will grab your attention. That 'spooky' sensation of turning your head to the side, almost without your will, only for you to lock eyes with someone you hadn't realised was watching you – it happens because your brain is continually monitoring far more sensory data than you can (or need to) become conscious of. If it spots on the periphery a potential survival threat – a gorilla, perhaps, or what it suspects could be a pair of watching eyes – well, that *demands* better data, and it's something that conscious awareness could help you to cope with, so, bang, your attention is grabbed.

Most visual illusions and other perceptual mistakes, constructions or omissions are common to all of us. But since our life experiences aren't all the same, neither are our expectations. And this can drive individual differences in what is perceived in even everyday situations.

No doubt, you have had an experience in which your personalised hallucinations have been thrown into stark relief by reality. I know I have. One Sunday morning in summer, I woke at about 4.30am, feeling hot. I got up to switch on the fan. As I got back into bed, in the dim light coming through the gap in the curtains, I saw my husband's hand sticking out of the crumpled duvet. Five minutes later, he walked into the bedroom, explaining that he'd fallen asleep on the sofa downstairs and had just woken up. It's not as though every morning, I see his hand in that position. This happened because I fully expected him to be there, and when I couldn't see his dark hair, the half-light (which made for 'fuzzy' sensory data, and so a heavier reliance on expectation) helped me to hallucinate a body part that is much closer in colour tone to our bedding. Somebody else standing in our bedroom wouldn't have seen what I did.

Most of the time, we don't experience this kind of extreme personalised hallucination, and we can agree on what we're seeing. Since we have fundamentally the same sense organs and brains and live on the same planet, one person's reality will usually at least roughly align with another. You might perceive a table as being redder than I perceive it. But we'd both agree that it's a table. Occasionally, though, we see things so differently that sparks fly.

#thedress is a case in point. This photo of a dress that went viral on social media because some people saw it as being blue and black while others argued – often passionately, sometimes angrily – that it was obviously white and gold was jumped on by vision scientists and psychologists. Before this, most had assumed that people with healthy colour vision see pretty much the same colours. A special issue of the *Journal of Vision*, published in 2017, was dedicated to research on it.[30] The explanation for the dispute, it seems, is that the brains of the people who saw it as blue and black were automatically assuming that it was being held up in indoor light, while the others were unconsciously assuming outdoor lighting.

Why should one person's brain assume one thing, and another's something else? It's thought that perhaps the 'indoor light' group had spent more time inside as children, when the visual processing system is very plastic, while the 'outdoor light' people had spent more time outdoors, and this early experience then influenced their perceptions into adulthood.

This active, constructive, flexible approach to perception has led some researchers, including Seth, to dub our experience of reality a 'controlled hallucination', a phrase he first heard used by the eminent cognitive scientist Chris Frith.[31] Each of us is absorbed in our own controlled hallucination, and lives in our own 'perceptual bubble', often assuming that everyone else sees things the same way – unless and until we come up against someone with a strikingly different reality.

It's lunchtime on a blustery winter's day, and I'm hurrying as fast as I can into Liverpool's Catholic cathedral. At once, I'm struck by the sound. Someone is tuning the grand organ. Loud, low, slow, bass notes are followed by a stop-start ascent through octave after octave, all the way to the 4,565th pipe. After a few moments of echoing silence, the cathedral's cavernous interior is filled with squeaky-high notes. Suddenly, the organist tumbles back down through the octaves.

For me, the absence of a melody – not to mention the low-frequency sound from the pipes – makes it feel confrontational, even disturbing. For Fiona Torrance, who lives in Liverpool and who chose this spot to meet, the sound experience is very different: 'I can see it in my mind's eye. It has a shape, and it moves. It's tubular-shaped, and the colours are changing. It was red, but as it gets deeper, it's turning purple.'

By the time she was about seven years old, Fiona had realised that she doesn't perceive the world the same way that most people do. It was only when she was in her mid-thirties, though, when a friend suggested that she might have synaesthesia, that she went for an academic evaluation. This confirmed that in fact, she has a whole collection of them.

Synaesthesia is a vivid demonstration of the huge role our brains play in creating the 'reality' around us. It is often described as a 'blending' of the senses. But the original Greek terms –*syn* – together, and *aisthesis* – sensation – are more accurate. For Fiona, the sounds of musical notes do automatically generate images of shapes and colours, a clear case of perceptions in one sense (vision) triggering perceptions in another (hearing). For her, colours themselves also generate individual touch, taste and temperature sensations. But, in addition, Fiona has one of the most common forms of synaesthesia to be identified. It's known as the 'grapheme-colour' type, and it involves only vision: 'I see colours for letters and numbers,' she explains. 'Though also for words that come out of people's mouths . . .'

Exactly how many types of synaesthesia there are is not clear, but dozens have been documented.[32] They include everything from letter/number – colour (this grapheme-colour synaesthesia is also the best-studied) to word – taste (for 'lexical-gustatory' synaesthesthes, these associations can be very specific; for one, the word 'jail', for example, generates the taste of cold, hard bacon, while 'tambourine' is a crumbly biscuit).[33]

To qualify as a synaesthete, a person must demonstrate consistent associations. For example, someone who says that for them the letter 'P' is light blue while 'S' is maroon must consistently match light blue to P and S to maroon on multiple tests at least 80 per cent of the time. Non-synaesthetes asked to come up with pairings don't show anything close to this level of consistency. Another hallmark of synaesthesia is that the pairings are effortless, and also, they are generally idiosyncratic.

Once considered to be rare, synaesthesia is now known to be relatively common. One recent study suggests it affects at least 4.5 per cent of people, meaning there could be 307 million synaesthetes across the world.[34] That's the entire population of the US.

So how do synaesthesias develop? Why does Fiona see things that I don't?

Since the early nineteenth century, it's been known that there's a hereditary aspect to synaesthesia. Recent research has confirmed that what's inherited is not a specific synaesthesia, but rather a propensity to develop one of some sort.[35] It's also become clear that synaesthesia develops early in life. Julia Simner at the University of Sussex has found that in young synaesthetic children, the associations tend to be quite chaotic, only settling down with age.[36]

For a child who has synaesthesia, the evidence suggests that it's theirs for good. Certainly, there's good evidence that synaesthetic pairings can withstand even temporary suppressions. Kevin Mitchell, a neuroscientist at Trinity College Dublin, has studied two people who have lost their synaesthesia for periods of time in their lives.[37] One was an unlucky young woman whose synaesthesia was temporarily suppressed by, among other things, a bout of viral meningitis, a concussion, and being struck by lightning.

For Mitchell, the main lesson from these case studies is that when a synaesthesia is established, while it can be temporarily affected by biochemical changes in the brain, it's remarkably stable over time. This suggests that once synaesthetic associations are set up and then consolidated, presumably in childhood, they remain 'hard-wired' afterwards. So how are they set up? And are there benefits to being able to see colour where it doesn't exist, or taste words?

One theory is that, for synaesthetes, neighbouring parts of the cortex that don't normally communicate do talk to each other, or they talk more than is typical. This idiosyncratic 'hyper-connectivity' could drive the unusual cross-perceptions. Individual variations in brain development, and also environment, may determine which kinds of cross-talk, and which synaesthesias, develop.

However, Jamie Ward, another leading synaesthesia researcher based at the University of Sussex, isn't convinced by this theory. In 2017, a team led by Simner reported that, as is the case for people diagnosed with autism, synaesthetes are more likely to have sensory sensitivities: they tend to perceive a light as being brighter than someone else does, for example, or a sound as being louder.[38] And the more synaesthesias a person has, the higher they score on the sensory sensitivity scale: 'If you have two types of synaesthesia, you'll score lower on the scale than someone with three, irrespective of what types they are,' Ward says.

Ward thinks that the theory that synaesthesia results from quirks of abnormal brain connectivity is wrong. Rather, he thinks that it stems from a drive, common to all developing brains, to maximise sensitivity to changes in sensory signals – to get the finest possible grasp on changes in the environment. Some young brains, though, are more 'plastic' than others. In them, this drive generates greater sensory sensitivities but also a level of instability in the system, allowing for neuronal connections between regions that don't typically communicate.[39] In childhood, this pattern of cross-talk is in flux. But as the brain loses plasticity with age, the synaesthetic pairings become 'set'.

In support of this theory, Ward points to various strands of evidence, including the results of his own computational modelling experiments, and also a 2018 study of people with grapheme–colour synaesthesia.[40] This study revealed unusual grey matter connections not just between regions that process letters and that process colours, but among other parts of the brain, too. 'A lot of things are not where they're expected to be,' Ward comments.

The superior memory performance of grapheme–colour synaesthetes[41] might be down to greater brain plasticity, which makes for easier learning, he thinks. Take the female synaesthete with the incredibly unlucky medical history, whom I mentioned earlier. She can't read sheet music. However, she can play the tin whistle, flute, glockenspiel, marimba and piano by ear. Her synaesthesia aids in all this, Mitchell reports, because 'wrong' colours flag incorrect notes. Fiona Torrance is learning to play the harp, and she reports benefiting from the same phenomenon: seeing the colours of the tones coming from her harp helps her to master new pieces.

People with more synaesthesias do also tend to score higher on the Autism Spectrum Quotient, a screening tool that assesses various autistic traits. Synaesthetes do not typically have the difficulties with social communication that are fundamental to autism; the overlap between the two groups seems to lie mostly in a trait called 'heightened attention to detail'. This suggests, that, as for autistic people, the perceptual representations of the world that are created in the brains of synaesthetes are focused more on the building blocks of a sensory scene (whether it's a painting, a city street, a sonata or the words coming out of someone's mouth) rather than on the 'big picture' of what all that detail together comes to mean. This sensitivity to detail could help to explain why those with synesthesia can develop remarkable abilities.

About one in ten people with autism have a remarkable ability of some kind too. Darold Treffert, an American psychiatrist, who has specialised in the study of 'savant syndrome', has shown that a range of striking abilities – including being able to do instant multiplication, prime number identification, calendar calculation and perfect-perspective drawing, or having absolute pitch or an extraordinary memory for facts – are far more common among people with autism spectrum conditions.[42] One of the best-known – since he's written an autobiography – modern-day cases is a British man called Daniel Tammet, who can remember the digits of the number Pi to over 22,000 decimal places.

Simon Baron-Cohen, now head of the Autism Research Centre at Cambridge University, was responsible for diagnosing Tammet's autism, when Tammet was twenty-six. Along with autism, he has a suite of synaesthesias. Numbers occupy specific locations in his mind. But they also have characteristic colours, textures and shapes.[43] As Tammet describes it, sequences of numbers create mental 'landscapes', which he can move through with ease. When he's performing a calculation, the shapes of numbers combine, to produce a new shape – the answer.

Could autism plus synaesthesia help to explain savantism, in at least some cases? It seemed plausible to Baron-Cohen. After studying Tammet, he ran further studies, and he found that synaesthesia is almost three times as common in people with autism, compared with the general population.

Julia Simner and Jamie Ward worked on a follow-up research study with Baron-Cohen, Treffert, and also James Hughes at Sussex. They looked at the prevalence of grapheme-colour, in people with autism and savant skills, people with autism without savant skills, and people who fell into neither category. The team found a significantly higher rate of synaesthesia only among people with autism and savant skills. Synaesthesia, then, is not more common in autistic people – rather, it's more common in autistic people who develop a prodigious talent.[44]

There are a few possible explanations for this, the team thinks. Firstly, grapheme-colour could be enriching their memories, making extraordinary feats of memory more feasible. Alternatively, something more fundamental could be at work.

A heightened ability to identify patterns, and to spot shared regularities between different sets of information, is a trait that could develop from others, including sensory hyper-sensitivity and excellent attention to detail. That ability may possibly contribute to both savant syndrome and, independently, synaesthesia. What lessons, then, can we draw from those with remarkable abilities? To what extent can our own sight, and our pattern recognition, be trained?

It is now clear that if there's a part of the visual cortex that responds to something specific – such as faces, or the orientation of lines, or even, for adults who as kids who spent a lot of time unconsciously establishing a dedicated region through extensive game-playing, Pokémon characters[45] – through repeated experience, you can train it to get even faster and more discriminative.

We can become accomplished at quickly identifying other kinds of objects, too. Different regions of the brain, including the prefrontal cortex, seem to handle expert recognition of objects that don't have a dedicated visual response region – whether that's cars or Qianlong-dynasty porcelain vases. But just because you don't have a dedicated visual region doesn't mean you can't get faster at processing and recognising all kinds of visual images.

During the Second World War, Allied lives were saved thanks to a technique for doing exactly this. Before computers, psychologists who wanted to present study participants with an image for very brief, even subliminal, periods of time used something called a tachistoscope (from the Greek *tachys* – swift, and *skopion* – an instrument for viewing). An American psychologist and vision specialist called Samuel Renshaw realised that he could use tachistoscopes to train pilots to get faster at recognising enemy ships and aircraft.[46] When they were repeatedly shown pictures of these craft for very brief periods, they increasingly became better at identifying them after only the merest of glimpses. The technique worked so well that in 1955, Renshaw was awarded the Distinguished Public Service Award from the US Navy.

Pattern recognition is a lot easier when your eyes are in good working order. Personally, I'm still pretending that I don't need reading glasses. I view documents on my computer at 125 per cent, hold print books as far away as necessary, and use my phone's camera zoom function, or younger friends, to help with annoying small-print on restaurant menus. What I should do, of course, is go to the optician.

The reason that, at forty-six – a highly typical age for presbyopia to become apparent – I'm finding it harder to focus on nearby objects has, of course, to do with the lenses in my eyes. The lens grows in a strange way: throughout our lifetimes, as new cells form on the outer edges, the older cells get pushed inward, making the central region denser and stiffer.[47] The stiffer it becomes, the harder it is for the surrounding muscle to squash the lens into the rounder shape required for focusing on near objects. With advancing age, these muscles get weaker, too, compounding the problem. Though lens hardening can begin even in the early twenties, for most people it takes decades of cellular stuffing before it becomes a genuine concern.[48]

The single biggest risk factor for presbyopia is age. Anyone over thirty-five is at risk. But there are other kinds of eyesight problems that have been directly linked not to a person's age, but to their lifestyle, and even young children are being affected.

The Yangxi County Experimental Primary School, located on

the southwest coast of Guangdong Province in southern China, was recently the setting for an experimental classroom that was one of a kind.[49] It was positioned in a clear spot, away from trees or tall buildings, and so from shadows. The supportive pillars and the crossbeams were made of steel. But the four walls and the roof were glass – clear for the bottom metre of each wall, topped with light-diffusing panes, which both cut glare and also screened the outside world, protecting the children from potential distractions. The entire point of the design was to allow in as much natural light as possible. The aim was to protect the children's vision.

People all over the planet are experiencing a 'state of mismatch' between the way our senses evolved and our current surroundings, argues Kara Hoover, an anthropologist at the University of Alaska, Fairbanks. And when it comes to vision, that mismatch couldn't be any starker. From an animal that spent virtually all its waking hours outside, many of us are now holed up in homes and offices, reading screens and books, under artificial light.

Evidence that this lifestyle change is taking its toll comes from skyrocketing rates of myopia – short-sightedness. For people with myopia, distant objects are blurry. This is caused by a slight elong-ation of the eyeball, which means that light from far objects is focused slightly in front of the retina, rather than on it.

According to some estimates, rates of myopia in the US and Europe have doubled in the past fifty years.[50] In East Asia, an esti-mated 70 to 90 per cent of teenagers and young adults are now short-sighted. In some countries, the rates of myopia are simply extraordinary. If you're a nineteen-year-old man living in Seoul, South Korea, who's not short-sighted, you're in a tiny minority – only 3.5 per cent of that population is so lucky. In China, 600–700 million of the total population of about 1.4 billion are myopic and need glasses – but many do not get them, particularly in rural areas.

Myopia is known to have a genetic component. But the recent, stratospheric escalation in cases has happened too rapidly to be explained by genetic change. Clearly, something environmental is involved. The question has been, what, *exactly*? Long periods of time spent poring over school books and focusing on screens is often blamed. The associations can certainly seem strong. Children living in Europe today spend a lot more time studying than they did in the 1920s, for example. In Shanghai, the average fifteen-year-old spends fourteen hours a week on homework, compared with five hours in the US and six in the UK. When it comes to screens, though, as myopia researcher Ian Morgan at the Australian National University points out, places like Taiwan, Hong Kong and Singapore already had myopia epidemics by the 1980s, when screen use was minimal.

In fact, careful work following children in the US and Australia suggests that it's not the number of hours that a child spends with books or screens that most increases myopia risk, but simply the number of hours that they spend indoors.[51] Time spent outdoors, in bright light, is key, argue a growing number of researchers, including Morgan. To be protected against myopia, Morgan estimates that children need to spend about three hours a day at light levels of at least 10,000 lux.

For many children living in sunny Australia, where only about 30 per cent of seventeen-year-olds are myopic, this is not a problem. On a sunny day, light intensities can soar to between 100,000 and 200,000 lux.

Ten thousand lux is about what someone in the shade, wearing sunglasses, on a bright day in Brisbane – or London – would enjoy. On an overcast day, however, light levels can drop to around 1,000 to 2,000 lux. But even that weak offering will still beat conditions in a classroom, in which 300 to 500 lux is typical internationally.

Morgan was involved in the glass classroom study at Yangxi County Experimental Primary School. This preliminary project showed that the classroom was practical – kids and teachers enjoyed being in it, and they could still read easily in such light.

The next step is to see whether glass classrooms can make a real difference to myopia rates, and Morgan is involved in planning this research. In the meantime, the best advice seems to be to have children outside as much as possible. This may not eliminate their myopia risk, above and beyond whatever inherited risk they might have, but preliminary studies in China and Taiwan suggest that just ensuring schoolchildren spend their break times outside, not inside, can make a difference.[52] For his part, Morgan argues that the first few years of primary school should be indoor study for half of the time and the other half should be outdoor-based activities.

What else can you do to protect your vision? Regular exercise reduces the risk of cataracts (clouded lenses), especially as we age.[53] It also cuts the risk of macular degeneration – the loss of cells in the region in and around the fovea – the leading cause of blindness in developed countries.

Diet, too, is important. The idea that eating lots of carrots allows you to see well in the dark is in fact an overly successful bit of *mis*information issued by the British Ministry of Information during the Second World War.[54] In an attempt to conceal the development of a new airborne radar, it publicly ascribed British pilots' ability to shoot down German bombers during black-out night-time attacks to the consumption of an excess of carrots . . . And it caught on. However, consuming foodstuffs that enable an adequate supply of vitamin A, such as carrots, broccoli, spinach and kale, is essential for the formation of rhodopsin, the light-sensitive protein in rod cells, and for the normal function of cone cells.

In extreme cases, however, a bad diet can cause blindness. In 2019, just such a case study of a British teenager hit national head-lines. This boy's junk-food diet of 'chips, crisps, white bread and some processed pork' had essentially caused his optic nerve to wither to the extent that, by the age of seventeen, he'd lost his sight.[55] The ophthalmologists at Bristol Medical School who evaluated him noted that one of the many nutrients that he was particularly deficient in was B12. People who follow a vegan diet and who don't supplement adequately are at risk of this deficiency, too, they cautioned.

Time outdoors, regular exercise and a healthy diet can all protect your sight. But there's also evidence that perfectly acceptable vision can be tweaked, to be better.

In 1999, the year that Debi Roberson's colour work on the Berinmo was published, a vision biologist called Anna Gislén at the University of Lund, Sweden, went on a research trip, with her six-year-old daughter, to Thailand. She wanted to study the 'sea gypsy' children, who live on the coast of and islands off western Thailand. Though our human eyes are poorly adapted for underwater vision, Gislén had heard that these children could easily collect small objects, such as clams, shells and sea cucumbers, from the sea floor. If this were true, how were they managing it?

Gislén tested children from a group known as the Moken. She placed cards showing various patterns under water, and found they were about twice as good as European children at discriminating between these patterns. On land, both groups performed about the same.[56] Clearly, there was something special about underwater vision in the Moken children, but whatever this advantage was, it didn't persist above the surface of the water.

Gislén reasoned that there could be two ways that these children might be improving their vision while in the sea. They could be shrinking their pupils to the tiniest possible size, which would increase their depth of field. Or they could be overcoming the brain's typical failure to bother to change the shape of the lens under water (because everything is just so blurry), allowing them to bring the patterns into focus. What she found was that, in fact, they do both.

In a further study, Gislén found that European children could be brought to the level of the Moken children after just eleven training sessions in an outdoor pool over one month. Their brains unconsciously learned to shrink their pupils and change their lens shape automatically. When tested again eight months after the last training session, these kids still had the same underwater acuity as the Moken.[57]

As we get older, our lenses become less flexible. Learning to alter the shape of the lens under water automatically is probably only possible for children. (Moken adults don't show this ability. They also tend to fish from the surface.) But demonstrating that lifestyle can change something as fundamental as our ability to see clearly was also a big surprise to vision researchers.

Still, it's not only children who can benefit from vision training. For adults, there are cases of training not the way the eye functions, but the way the brain processes incoming visual information – with astonishing results.

Sue Barry was born with what today doctors call a 'squint', but at the time was described as cross-eyed. Surgery on the muscles of her eyes during childhood did correct this to a large extent. But because she couldn't properly coordinate her eyes to work together, she was denied the slight differences in information from the eyes that allow most of us effortlessly to perceive depth.

To her, the world looked flat. 'I would see myself *on* the pane of glass of the mirror,' she explains. 'If there was a speck on the mirror, I'd think it was on myself.' In a remarkable demonstration of just how 'blinding' our perceptual bubbles can be, it was only when Barry took a course in neurophysiology at university that she realised that other people saw things very differently.

It's very hard for anyone who has grown up seeing in three dimensions to understand how she saw things. Simply putting one hand over an eye won't change an awful lot. (You can try it.) 'Yes, there might be a very subtle difference, but your brain is using all the experiences it's had to recreate a sort of 3D world – and so it's a very different experience,' Barry says.

Once she comprehended, as far as she could, what she was lacking, she also realised something else: that she'd never see the world in three dimensions. The received wisdom was that the critical period of sensitivity for developing 'binocular cells' in the visual cortex closes after the first few years of life. The adult brain, it was argued, is just not capable of the plasticity required for this kind of radical change.

And yet . . . well, here is her own, personal testimony to just such a transformation.

In her late forties, Barry found it increasingly hard to see objects in the distance. She consulted an optometrist, who informed her that she was using each eye, individually and alternately, to see anything other than objects within a few inches of her face. The optometrist prescribed a series of training exercises that were designed to help her to use her eyes in concert, with the hope that she'd start to fuse the images from each into one. Within a few weeks, Barry noticed extraordinary changes in how she saw her own home: the edges of the light fitting in her kitchen seemed to be more rounded, and she felt, for the first time, that it occupied a space between her and the ceiling.

The odd experiences kept on coming. As Barry wrote excitedly to the neurologist Oliver Sacks (whom she'd met at a space shuttle launch party), the steering wheel in her car suddenly 'popped out' from the dashboard. Leaves on bushes seemed to stand out in their own little spaces. The head of a complete horse skeleton in the basement of her work building seemed to stick out so much, she jumped back and cried out.[58]

Evaluations of Barry's vision confirmed that, yes, she was now seeing in 3D. And her experience is not unique. She has since gathered accounts of even more rapid transformations. None sounds more unlikely than the case of Bruce Bridgeman, professor of psychology and psychobiology at the University of California, Santa Cruz.

Like Barry, Bridgeman grew up without stereo vision. In 2012, at the age of sixty-seven, he went to watch the movie *Hugo* in 3D, and he put on the appropriate glasses. At first, as he'd anticipated, the film looked flat. But suddenly, it seemed to pop. After he left the cinema, his new-found ability to see depth stayed with him. It's not just him, Barry says; other people have said that all of a sudden, they saw in three dimensions.

In 2017, Barry and Bridgeman published the results of a questionnaire-based study, in which they asked other people who had developed 3D vision in adulthood what this was like.[59] More than one-third described their depth perception as 'astonishing'. Seeing in stereo gave them a qualitatively different sense, Barry reports: 'a sense of the volumes of space between things'.

This is still a developing field, and exactly what's happened in the brains of these adults isn't clear. Barry suspects that a 'switch' to seeing the world in 3D typically occurs during babyhood, but can happen in adulthood, too. However, it's possible that perhaps she did have a few early 3D experiences – enough, anyway, to allow some binocular cells to develop, enabling her to develop proper depth perception later in life.

Certainly, the idea that someone who has never seen in 3D can learn to do it in adulthood is highly controversial. A few days after speaking to Barry, I found myself talking to an ophthalmologist at a friend's barbecue. She performs corrective surgery for just the type of problem that Barry had at birth, and she stressed that she wished, for the sake of her patients, that adult development of 3D vision in someone who's never had it were indeed possible. However, it may soon be the case that for those of us with more standard vision, transformations that are just as dramatic are in store.

In 2016, an Australian-led team reported that they had developed nanocrystals that could receive and concentrate infrared heat radiation and convert it into light that we can see.[60] In theory, these crystals could be incorporated in glasses, to make lightweight night-vision specs (primarily for military use, though the team also suggested their potential for enabling night-time golf . . .) In 2019, researchers at the University of Massachusetts Medical School went a step further, reporting that they had injected nanoparticles with essentially the same function into the retinas of mice.[61] The particles converted infrared light into shorter wavelength, green light. The team used brain scans and light tests to confirm that the mice really could detect and respond to infrared light. Even in daylight, they could spot infrared patterns.

The team envisages uses for animals: 'If we had a super dog that could see near infrared light, we could project a pattern onto a lawbreaker's body from a distance, and the dog could catch them without disturbing other people,' suggests Gang Han, the principal investigator. But there's no theoretical reason that it shouldn't work for humans, too. 'When we look at the universe, we see only visible light,' notes Han. 'But if we had near-infrared vision, we could see the universe in a whole new way. We might be able to do infrared astronomy with the naked eye, or have night vision without bulky equipment.'[62]

It's not beyond the realms of possibility that we might be able to swap our human eyes for a bionic, better version. In 2020, a team at Hong Kong University unveiled a 3D bionic eye that copies the structure of a natural eye, but which can be tweaked to offer sharper vision and extra functions, such as infrared night vision.[63] The eye is sometimes held up as an example to counter claims of 'intelligent design' (which omniscient being would design the eye to have nerve fibres bundling through the retina, so creating a blindspot?) The new eye, with its nanowire light sensors, wouldn't require this, eliminating the blindspot. It *would* be intelligently designed. It's still early days for this technology, though. For now, the team stresses, current bionic eyes are still no match for their natural counterparts.

For Aristotle, sight was the sense that provides the richest information about the outside world. This idea has been expanded in our everyday language. A discovery may 'throw new light' on a subject, or 'illuminate' it. If I suddenly grasp what someone else is trying to convey, I might say, 'I see what you mean.' It's possible to 'lose sight' of a goal, and even have 'dark' thoughts – but ultimately to 'see the light'. What started out as a very simple receptor for registering light versus dark has become a route not only to recognising food and family but also to conveying our deepest feelings. It's also become, of course, a conduit to aesthetic pleasure. For humans, sight is not only of practical use. We pay good money – astronomical amounts, in some cases – to gaze upon a beautiful thing.

Still, in his classic 1961 book, *Art and Illusion*, renowned art historian Ernst Gombrich wrote: 'It is the power of expectation . . . that moulds what we see in life no less than in art.'

The 'power of expectation' . . . discourse over the years in both the arts and scientific fields has resisted, or at least avoided, this thesis. As Anil Seth has explored in his own work,[64] the idea that it's important to consider what each individual person brings to the perception of an artwork became unfashionable in art history. In psychology, meanwhile, research focused on the fundamental neuroscience of vision.

However, as we have just seen, the latest work on how our brains handle signals from our eyes is – well it's enlightening, for sure. Decades ago, Gombrich knew that seeing is about so much more than processing what's 'out there'. This insight is now gaining much more academic attention, and it marks a fundamental shift in our understanding of not only sight, but also our other senses.

In the seventeenth century, the French philosopher and scientist René Descartes argued that the senses could not be trusted, and were, therefore, unreliable sources of knowledge about the world. In so doing, he put himself in direct opposition to the Aristotelian establishment, which held that the senses are our only source of accurate knowledge. But clearly, if there is such a thing as full, true, objective information about the world out there, that's not what you perceive. Equally clearly, your brain couldn't handle that truth.

Your fundamental drives are to survive, reproduce and thrive. Your senses deal in only the types of information that help you – a human on planet Earth – achieve these things, and your brain, locked away in its skull, has to use all its resources to fathom, as quickly as possible, the meaning of an unrelenting torrent of incoming sensory signals. Memories of whether or not the other side of the bed is typically occupied, or of what variations in light intensity typically do to colours, are all usually excellent resources, speeding that process along. But these leg-ups to fast perception make us vulnerable. As we now know, it's perfectly possible to see things that aren't actually there – as well as not to see things that are.

As we've also discovered, research to better understand when and how this happens – and why some people's visual perceptions are more fallible than others – holds lessons for all of us. While this is certainly true for vision, it's also the case for the sense that we'll come to next: hearing.

2

Hearing

Why 'Dancing Queen' sounds different in Bolivia

> That which sounds . . . is that which produces motion in such
> air as is one in continuity up to a hearing-organ.
>
> Aristotle, *De Anima*

Aristotle was right, of course. For you to hear something, sound waves must make contact with your hearing organ – your ear.

All kinds of life forms, from plants to earthworms, can sense types of vibrations that we would perceive as sounds. Well, earthworms can sense some. Charles Darwin, who was famously fascinated by worm senses, experimented with them by shouting at them, blowing a metal whistle and asking his son to play his bassoon loudly, all the while observing them closely. The worms didn't react to noise – until he put them, in their pots, on top of the piano. Then, the powerful vibrations sent them instantly retreating into their burrows for cover.

Plants, too, can sense what we'd call sounds.[1] The small flowering plant *Arabidopsis* can even distinguish between recordings of the sound of wind, a caterpillar munching on a leaf, and the mating song of a leafhopper. The leafhopper song was chosen for these tests, run at the University of Missouri, because it has a similar frequency to the caterpillar's chewing sounds. But the plants were not duped. They only produced more mustard oils, which caterpillars find off-putting, in response to the sound of munching.[2] It's a remarkable demonstration of the sensing abilities of plants. However, since neither earthworms nor plants have ears or brains, they don't 'hear'.

Our sense of hearing is a physical sense, and it's related to touch.

A movement – whether that of someone else's vocal cords or a herd of elephants stomping towards you – triggers waves of energy through the surrounding molecules of air, or liquid, or even a solid. We can sense some of these types of waves using 'mechanoreceptors', which respond to physical force, located in various parts of our bodies. Think of how you can *feel* a church organ deep inside you, or how a particularly loud bass can grab at your chest. These perceptions depend on the stimulation of receptors that we'll come to in Chapter 5, which looks at touch.

But deep within your inner ear is an organ containing a particular cluster of mechanoreceptors that evolutionary time has tuned to be exquisitely sensitive to the wave frequencies produced by the kinds of movements that matter most to us. Our alchemical brains transmute these base, physical energy waves into something phenom-enologically special – sounds. We don't *detect* sounds. We detect a vibration; sounds are what we make in our heads.

That famous philosophical thought experiment: 'If a tree falls in a forest and there's no one around to hear it, does it make a sound?' has a very clear answer, then: no. Well, not unless there's another animal around whose brain is set up, like ours, to perceive the resulting air oscillations as sounds. And of course, there are many.

Our sense of hearing enables us to read our environment at a distance. It provides potentially life-critical information about where we are, and what's around us, even if it's too far away for us to see, or it's obscured, or it's night-time – or we're asleep. It also allows for a form of easy communication between us and other animals. These are advantages that have obvious benefits for the wide variety of life forms that do undeniably have ears. Take sharks.

Sound waves travel easily through water, making them a more useful source of information than light. Though sharks are renowned for their electrical sense and for their sense of smell, they also have an acute sense of hearing. A shark can detect the distinctive low-frequency sounds made by a sick or wounded animal, or a flailing swimmer, from hundreds of metres away. However, it does this with an ear that's not quite like ours. In fact, the evolutionary route to our mammalian ear is so improbable, it reads almost like something from the pages of a Marvel comic.[3]

The shark lineage and ours diverged about 420 million years ago. The path to sharks leads one way, via other, earlier cartilaginous fish. The path to mammals goes via bony fish and early synapsids, a group of animals with some mammal-like characteristics. Roughly 230 million years ago, our synapsid ancestors evolved an extra pair of bones in the jaw. No one knows why.

In the next, even stranger chapter of the story, two now-redundant jaw bones gradually became smaller and moved into the ear, where they became the malleus (hammer) and incus (anvil). These are two of the three tiny bones in our middle ear that amplify incoming sound waves. Sharks (and reptiles and birds) have just one: the stapes. Because the three bones working together transmit sound more efficiently than one, they make for more acute hearing and a greater sensitivity to high-pitched sounds.[4] So if a shark were to bite you, it would sense your panicked thrashing, but it would not hear your screams.

What can you hear right now? Whether it's the screech of the brakes of an underground train or the drumming of rain on the roof, it's because incoming sound waves are vibrating your tympanic membrane (your eardrum), which separates the external, sound-funnelling part of your ear from your middle ear.

As the eardrum vibrates, it in turn sends tremors through the malleus, incus and stapes, which then transmit them as pressure waves to your liquid-filled cochlea, in the inner ear. The cochlea (from the Greek *kokhlias* – snail) is a hollow spiral of bone, measuring, on average, 8.75 mm long and 3.26 mm high. It twists like an enclosed helter-skelter slide, the membranes and cells critical for hearing coiling through it.[5] Embedded in the top of the basilar (base-level) membrane are the auditory hair cells of the organ of Corti. Their microscopic hair-like 'stereocilia' stick up, the tallest of them protruding into another, gel-like membrane above. Incoming pressure waves make the basilar membrane ripple. Different regions of it respond more to certain frequencies than others. The stiffer region near the base of the cochlea moves most in response to high-frequency – high-pitched – sounds, like a squeal. Regions near the top respond most to lower-frequency sounds, like drumming.

When a region of the basilar membrane vibrates, the stereocilia are pushed up and bent against the gel-like membrane above. The mechanical force opens up tiny channels, of a type known as ion channels, in filaments that connect the stereocilia. Potassium ions then flood in through those channels. This initiates an electrical signal, which is carried by the auditory nerve to the brain – with the end result that you perceive a sound.

Your brain first processes a sound's frequency (how high or low it is), its duration and intensity. Sound information also goes on to other regions of the brain that allow you to localise the noise, and that prepare you for a physical response, like talking back, or running away. Further processing in the auditory cortex allows you to make proper sense of the signals: recognising that coo-ing as coming from a pigeon, say, or that song as the one your ex played over and over in the weeks before you broke up. Or, for a newborn baby, even, the particular sound pattern of its mother's voice.

Before you are born, your sense of hearing is up and running. Starting at around twenty weeks of gestation, foetuses respond to a loud noise with a reflex scrunching of their eyes. (They would blink, if they were able, but their eyelids won't open for about another six weeks.) At around the same time as the ear canals open up, allowing sounds in, connections form between the thalamus (that relay station for incoming sensory data) and the sensory processing regions of the cortex. Even before the third trimester, then, the cochlea is open for business. Though the uterus, not to mention the rest of the mother's body, muffles noises, we come into the world pre-packaged with a familiarity with the sounds of our own home, language and culture.

We know this in part thanks to an experiment conducted in 1980, involving ten new mothers.[6] The study is now regarded as ground-breaking, but to those mothers, it must have seemed pretty strange.

Shortly after giving birth, the women were asked to record a story by Dr Seuss, *And To Think That I Saw it On Mulberry Street*. At some point in the next twenty-four hours, their newborns were laid down in their bassinets, earphones were placed over their ears and a plastic nipple that released no milk was placed in their mouths.

The nipple was hooked up to a computer that itself was linked to a speaker. The psychologists behind the study set up the kit so that if a baby sucked in a certain way (with either short or long pauses between bursts of sucking, depending on the experimenter's preference), their own mother's voice recording would be played, while a different sucking pace would trigger a recording made by another mother.

One infant in the study seemed actively to prefer not to hear its mother, and one didn't seem to mind whom it heard. But most sucked in such a way as to be able to hear their own mother's voice. When the psychologists then reversed the link for each infant between sucking pace and voice, a few even learned to switch their sucking style in order to hear their mother talking again.

Before this experiment, few psychologists were interested in the behaviour of newborns. But this work really piqued others' curiosity. The finding on the babies' rapid learning ability aside, what it hinted at very clearly was that we come into the world knowing our mother's voice. Future research by the same psychologists confirmed that hearing in the womb, rather than in the hours after birth, is responsible. Knowing what we do now about the development of hearing, it's not a big surprise: by the time they were born, the babies had been listening to their mother for months.

At birth, if all these babies were typical, their cochleas contained around 16,000 of those sensory hair cells. (That might sound a lot, but contrast that figure with the roughly 100 million light-sensitive cells in your retina.) But even if you were born with precisely the same number of auditory hair cells as me, chances are we don't hear the world in the same way.

Our hearing sensitivity – whether you can't bear a gentle tapping or are quite comfortable with your neighbour's new drum kit – is certainly influenced by our genes. Mutations in about seventy genes have been linked to hearing loss and deafness, and, compared with the influence of genes on variations in some other types of sensing, the heritability of hearing sensitivity is high. According to one study of healthy Finnish twins, for example, 40 per cent of the variation in sound sensitivity observed in this group could be ascribed to genetic differences.[7]

This genetic variation may help to explain why we have such different discomfort levels for everyday noises. Some sounds are just awful, no matter how loud they are. A human scream, for example, activates brain regions involved in aversion and pain (recent work showed that fast but perceptible fluctuations in loudness, which characterise screams as well as some other animal alarm calls, and also police sirens, drives this).[8] But when it comes to traffic noise or a lawnmower, say, it's generally the volume that determines whether it's perceived as hideous or not. However, what people with healthy hearing consider to be an 'uncomfortable' volume can vary by around twenty decibels.

Bear in mind that the decibel scale is logarithmic. Most people perceive a sound that is just ten decibels louder than another to be twice as loud. What is or isn't 'too loud' is, then, hugely subjective. A lawnmower, powering out roughly ninety decibels of noise, or a leaf-blower, could be hideously intrusive for one person, whereas for their neighbour, or partner, it may not be a problem at all. (How many neighbourly spats, I wonder, have been triggered by arguments about just how loud something is, each person trapped in their own perceptual bubble, convinced that they are right . . .)

But differences in what and how we hear go far beyond volume, as we'll find out – and lead us to big questions about what makes us human.

Music-making is profoundly human. All cultures make music. And compared with other primates, our brains do seem to be especially tuned to respond both to musical beat[9] and to pitch in music and speech. Though your brain and a macaque monkey's brain handle vision in much the same way (which suggests that its visual experience of the world is probably similar to yours), there are regions of your auditory cortex that are especially responsive to tones, rather than to toneless noise. The macaque brain doesn't show this division.[10] 'These results suggest the macaque monkey may experience music and other sounds differently,' says Bevil Conway, a neuroscientist and sense specialist at the US National Institutes of Health, who was involved in this study. He adds: 'The results raise the possibility that these sounds, which are embedded in speech and music, may have shaped the basic organisation of the [human] brain.'

Though, unlike other primates, all people make music, we don't make it in quite the same way, however. Traditionally Western music – whether it's Beethoven's Fifth or ABBA's 'Dancing Queen' – is based on octaves. Within this system, the pitch of a note doubles with ascending octaves. So notes with the frequencies of 27.5 Hz, 55 Hz, 110 Hz, and so on, are all As, just in different octaves. Your vocal range might be significantly higher than mine, meaning that if you were asked to sing 'Dancing Queen', you'd start in a different octave than I would, but assuming we could stick to the equivalent notes, we could sing along to the music together.

The idea that these notes, in different octaves, are equivalent is so overwhelmingly dominant in the West that it's been impossible to know for sure whether it really is the product of biology – relating to the specific regions of the cochlea that are caused to vibrate – or just extensive experience with octave-based music. To explore this, you need to investigate responses among a culture that has had very little exposure to Western music. That isn't easy. But in 2019, just such a study was reported.[11]

The researchers studied the Tsimane, a group of people living in the remote Bolivian rainforest, mostly along the Amazon. Not only have the Tsimane had relatively little contact with Western culture, but they don't use the octave system in their own music. The team administered simple tests, and compared the Tsimane responses with those of a group of Americans, both trained musicians and non-musicians. Each participant was played tunes of only two or three notes, and asked to copy them by singing them back. Any given tune fell within just one of eight octaves, ranging from very low to very high, but the singer of course had to sing them back within their own, restricted range.

The results were clear. When the Americans – and especially the trained musicians – were played the three notes A-C-A, for example, they would sing back A-C-A, just in an octave appropriate to their own range, which might have been higher or lower. When the Tsimane were played A-C-A, they'd sing back a tune with a middle note that was appropriately different in pitch, compared with the other two. But the absolute pitch of these notes bore no relationship to the absolute pitch of the stimulus notes. So while they might have been played an A as the first note, they didn't sing back an A. This would be equivalent to you repeating the opening '*Oooh*' of 'Dancing Queen' as C#-B-A but me coming back with F#-E-D.[12]

This clearly implies that it's exposure to Western music that makes us hear various As (and Cs, and so on) as being equivalent – rather than anything to do with precisely which parts of the cochlea are being stimulated. It might seem intuitively obvious to you, or me, that an A in another octave is still an A. But this work shows that this sense of equivalence has been *learned* – it's caused by culture, not biology. And it's a powerful example of the power of culture in affecting what we hear.

Even more remarkable, though, is this: how our brains process music is informed not only by our previous experience of music, but also by the language that we speak.

There are cultural differences in the metaphors that we use to refer to pitch, and they can influence what we think. I've used the standard Western metaphors in the paragraphs above – describing an A at 110 Hz as being 'higher' than an A at 27.5 Hz, for example, and referring to some vocal ranges as being 'higher' than others. To an English-speaker it feels so appropriate. But not all languages talk of 'high' or 'low' notes. People who speak Farsi (which is mostly spoken in Iran) use 'thin' or 'thick' instead. The Kpelle of Liberia use 'light' or 'heavy'. The Suyá people of the Amazon basin use 'young' or 'old'.

I've ordered these alternatives to match the order of 'high' and 'low', but I'm sure I didn't need to – these different ways of referring to pitch make instinctive sense, too, don't they?

Still, Asifa Majid, at the University of York, who specialises in the study of cultural influences on perceptions, has found that these metaphors can affect what we hear. Until recently, Majid was based at a lab in the Netherlands, where 'high' and 'low' are commonly used. She asked groups of Dutch- and Farsi-speakers to listen to a tone and then sing it back. At the same time as they heard the tone, they were shown a line positioned on a screen. When that line was high up on the screen, the Dutch-speakers tended to sing the same note higher than when it was low down – but wherever the line was, it made no difference to the Farsi-speakers. For them, however, being shown a thick or a thin line influenced the pitch of the note that they sang back – but this made no difference to the Dutch-speakers.[13]

Majid then tried a version of this task with four-month-old Dutch babies. Of course, they couldn't sing notes. Instead, she looked at how much attention they paid to matching or mismatching note/line heights and thicknesses. She found that all the babies were sensitive to both line height and thickness. When listening to a low-pitched note (I just can't escape that metaphor), the babies preferred to see either a low or a thick line, rather than a high or a thin one. The same was true for a high-pitched note and a high or a thin line.[14] Majid concludes that we are born sensitive to both of these sound/spatial associations and metaphors – and others, no doubt – but as we grow up, associations that aren't explicit in our native language lose their influence over us, if not their intuitive sense.

Why should it seem obvious that higher pitched would be 'light' rather than 'heavy'? One explanation is that physically heavier animals (men versus women, elephants versus mice, stags versus fauns, and so on) make lower-pitched sounds as they move. This knowledge certainly influences our judgements about body size – and it can be hacked, with extraordinary results.

Ana Tajadura-Jiménez is a former telecommunications engineer turned psychoacoustics specialist at University College London. She's the brains behind a pair of slip-on sandals that she calls 'magic shoes'.

The sandals are fitted with microphones, which are linked to earphones. When someone walks in these sandals, the sound of their feet hitting the ground is picked up by the microphones. Before it's played on to them through the earphones, though, it is filtered, so that they hear only the higher-frequency portion.

This has the effect of making the wearer *sound* as though they're physically lighter. And Tajadura-Jiménez has found that their brain registers this and adjusts its representation of their body accordingly, making them actually *feel* lighter and slimmer. In fact, people using the magic shoes also report feeling happier, she reports, and take on a more sprightly gait.[15] Tajadura-Jiménez hopes that this kind of self-deception could have advantages, as it might perhaps motivate people to do more physical activity.

Research using the magic shoes helps to show that what we hear can influence our perceptions of ourselves, as physical entities. In this case, when what we hear is manipulated, our sense of who we are warps – and we feel lighter. But what if, rather than sounds being distorted externally, your own brain couldn't process auditory signals properly? What might happen to your sense of who you are, and what you're experiencing?

There's plenty of evidence that, as with vision – when I hallucinated my husband's hand, for example – a reliance on expectations can even make us hear things that aren't there.

A lab study run at Yale University way back in the 1890s demonstrates this neatly. The participants in this study were repeatedly shown an image while simultaneously being played a tone. Soon, they reported hearing the tone whenever they saw the image, even when it wasn't played. They had become so conditioned to expect it – their expectations that it would happen were so strong – that even in the absence of appropriate sensory signals from their organs of Corti, they heard it. Their predictive brains had put too much weight on past expectations, over sensory evidence.

In your own life, perhaps your expectation of hearing certain sounds – such as your mobile phone ringing – is so strong that your brain occasionally hears them for you. If this happened all the time, life would be very confusing. To understand which of our sound perceptions are real, we need to be able to update and also sometimes dial back on expectation. But this, it seems, is easier for some people than for others.

In 2017, a Yale psychiatrist, Philip Corlett, and his colleagues, repeated the old image/tone experiment, but with four different groups of people: healthy people, people with psychosis (which entails a loss of contact with reality) who don't hear voices, people with schizophrenia (a form of psychosis) who do hear voices, and a fourth group, who regularly hear voices but who don't find these voices disturbing (a group that included people who regarded themselves as psychics).[16]

Everyone in the study was trained to associate a chequerboard image with a one-kilohertz, one-second-long tone. While imaging the participants' brains, the researchers then manipulated the presentations, so that sometimes a tone wasn't played. They found that the 'psychics' and the people with schizophrenia were nearly five times more likely to report hearing a tone when it wasn't actually there as the people in the other groups. What's more, when they did report hearing a non-existent tone, they were about 28 per cent more confident that they'd heard it than the others were when they made the same mistaken report.

The brain imaging showed that those with the strongest auditory hallucinations had altered levels of activity in a few different regions, including reduced activity in the cerebellum. The cerebellum, a bulging structure at the base of the brain, is involved in coordinating physical movements, a process that requires constant and accurate updates about the outside world. The findings suggest that the cerebellum plays a critical role in ensuring that sensory data are adequately taken into account when our brain generates its sound perceptions. If that doesn't happen properly, reality can drift.

In fact, some researchers think that to understand hearing voices in general, and schizophrenia, in particular, we need to take a much closer look at how the brain processes sound information, as well as other sensory signals. This work also holds important lessons for the rest of us. We so often undervalue our senses, and yet these systems run right at our very core. By taking a closer look at what happens when things go wrong, we can come to appreciate just how critical they are to our everyday understanding of not just the external world, but our own selves.

*

Starting when I was about 20 years old, I heard the voices of demons screaming at me, telling me that I was damned, that God hated me, and that I was going to hell . . . . The voices were so frightening and disruptive that much of the time I was unable to focus or concentrate on anything else.

To a point, they generally are anything but kind to me. They can be brutally sarcastic and intrusive.

I hear a mixture of men and women, but no children. They usually tell me to do things, but not dangerous things. Like they'll tell me to take out the garbage or check the lock on the window or call someone. Sometimes they comment on what I'm doing and whether I'm doing a good job or what I could be doing better.

These descriptions of voices that aren't real come from a survey led by Charles Fernyhough at the University of Durham.[17] Hearing voices is not uncommon. Results from around the world differ widely, but somewhere between 5 and 28 per cent of people know first hand what it's like. You certainly don't have to be diagnosed with schizophrenia or another form of psychosis to hear voices or other sounds, but about 80 per cent of people with schizophrenia report auditory hallucinations.[18] The recent Yale chequerboard/tone research suggests a reason for why this could happen. But beyond an apparent under-reliance on sensory signals, why might this be?

When you talk out loud, your brain immediately starts to suppress auditory processing of your own voice. This is why when you hear yourself on a recording, it can be a surprise to hear what you actually sound like. This automatic response helps your brain to be very clear about when you're speaking, or someone else is.

There is evidence that people with schizophrenia just don't do this automatic own-voice suppression very well.[19] In theory, notes neuroscientist John Foxe, at Rochester University, USA, this could lead to confusion: if your brain mistakenly assumes that someone else is talking, when in fact you are, or the voice is coming from inside your own head, its 'best guess' about what's going on – and so what you actually perceive – could easily be wrong.

Now recall that according to the predictive perception model, if sensory data aren't clear and precise (and you can't do something to enhance them), they will be relied upon less, and your predictions about what's happening will be weighted as being more important, skewing the resulting perception in expectation's favour. This could help to explain auditory hallucinations, but it might also explain much more.[20] Because for people with schizophrenia, there's growing evidence that their sensory data can be very unreliable indeed.

It's not only atypical sound processing that is being observed in people diagnosed with schizophrenia. Problems with interoception[21] (the sensing of bodily states) and proprioception[22] (sensing the position of body parts), as well as balance, are common, too. Some people with the disorder report delusions of control – feeling that someone or something else is driving their actions. As one individual described it: 'It is my hand and arm which move, and my fingers pick up the pen, but I don't control them. What they do is nothing to do with me.' These delusions have been ascribed to difficulties in perceiving their own muscular movements, while deficits in inner-sensing – a reduced sensitivity to signals from inside the body – are also thought to weaken the critical awareness of what's happening to them, as individual selves.

Over the past few decades, John Foxe has identified all kinds of sensory deficits in people with schizophrenia. One finding is that their brains don't adapt well to ongoing visual and touch information.[23] If, for example, you're wearing jeans right now, sections of the fabric will be constantly or repeatedly touching your legs, but once your jeans are on, you stop noticing this. However, when people with schizophrenia are repeatedly touched, their brain response stays strong. Foxe speculates – and he wants to stress that it is speculation – 'If you do not adapt to persistent stimulation, then it is easy to imagine how things that should fade into the background might continue to impinge on one's consciousness and lead to a disordered or warped sense of reality.'

Someone with disordered touch processing might give sensory signals such little weight, and expectations about what's likely to be happening such a strong emphasis, that they even feel things that just aren't there.

'Ms A' presented to emergency department (ED) with complaints of 'lice and bugs' crawling from her skin. She reported using several tubes of permethrin cream over the past 2 weeks with no relief of her symptoms. Initially she had only been itching, with a feeling of crawling under her skin, but upon presentation to the ED she reported that the bugs were crawling 'all over' her.

This is an extract of a case study of a woman with 'delusional bug syndrome',[24] which was first described in the medical literature in 1938. It entails a rigid, misguided belief that some kind of insect or parasite is crawling over or burrowing into and under the skin. It's usual for people with the syndrome to see a string of dermatologists, only for negative skin-test results to do nothing to dent their conviction that the infestation is real (making them, by definition, delusional). It's also often reported by people who go on to be diagnosed with schizophrenia.

There is also evidence that, like some autistic people, people with schizophrenia don't integrate visual and auditory information properly.[25] This specific impairment could make it hard for them to comprehend speech.

'One of the ways in which you and I understand each other is the words themselves,' Foxe says. 'But it's also the prosodic content, right? The lilt, the changes in frequency in my voice are telling you when I'm trying to emphasise a point or if I'm being ironic, and so on. Patients with schizophrenia are *terrible* at prosody. They have real trouble with the emotionality in a voice, or understanding if someone is asking a question or making a statement.'

This seems to be a high-level type of problem – they're just not understanding the non-verbal content of what someone else is saying. However, Foxe and his team have found that patients with schizophrenia with the worst prosody processing difficulties don't just have trouble hearing changes in tone in a voice, but trouble hearing tones, full stop:[26] 'When we just used simple tones – not fancy words, just tones – there was a very strong correlation with an inability to just hear basic frequency differences, which is of course a sensory processing disorder. If you can't hear frequency differences, you can't do prosody. So here's a case where something that has a very fundamental sensory explanation comes across as a higher order social, cognitive deficit.'

When people think about schizophrenia, Foxe says, the first things that come to mind are the paranoia, the delusions, the disorganised thinking. But there is mounting evidence that there is something fundamentally wrong with how these people hear, as well as how they feel touch, see the world, perceive signals from inside their own bodies – and so much more.

I ask Foxe if he feels ready to say that schizophrenia is a sensory disorder: 'Am I ready to say that definitively? . . . No. But it definitely has a sensory processing component to it, and it's much more prominent than people think. In every sensory modality, we find processing deficits. So you might think of it as being a cognitive disorder – that it's high-level thinking that's going down – but then you find all these sensory deficits. It's an interesting exercise to flip it around and say, "Okay, could you end up with these high-level thinking deficits by just having basic-level sensory processing deficits?" And that may very well be the case.'

Schizophrenia is an extreme example of a reality warp. But, clearly, there is evidence that (for people who can hear) the way our brains handle sound signals is important to our sense of self and our perceptions of reality. For most of us, however, our hearing is not static throughout our lives. Though sudden deafness is not common, the majority of us will suffer deteriorations in how we hear with age. And there is now evidence that this 'typical' age-related loss can have dramatic impacts on our mental and physical health.

An estimated one in three people aged between sixty-five and seventy-four have difficulty hearing. This figure rises to one in two for people over seventy-five. Being unable to hear friends or a doctor, or a doorbell or a smoke alarm properly can, as the US National Institutes of Health puts it, be 'frustrating, embarrassing, and even dangerous'. If you can't really hear your favourite songs any more, or birdsong in your garden or what your partner's saying to you, it can also be a source of sadness and, ultimately, isolation.

> Pamela (86) doesn't go down to the common areas in the residential home as she can't hear all the conversation – she finds it difficult in a group.
>
> Julie (70) finds group conversation difficult, mainly because of the hearing impairment.
>
> Colin (92) After you've gone it will be just like a morgue again.

These comments are from a British study that sought to go beyond the statistics, to explore how sensory loss can blight the lives of older people.[27]

The brains of people who are born without a sense, or lose it early in life, can adapt. There are many examples of this, including the remarkable case reported in 2017 of an Australian boy who, due to a rare disorder, lost his primary visual cortex (V1) shortly after he was born, but who when tested at the age of seven had near-normal sight (other brain regions had been roped in to take on V1's role).

Studies of people who were born blind, or who lost their vision early in life, have also revealed differences in the auditory cortex, compared with sighted people. In one recent study, the researchers linked these differences to being better at differentiating between sounds of slightly different frequencies, and to tracking the sounds of moving objects, such as cars or people.[28] The study included two people who had been blind since infancy, only to have their sight restored through surgery in adulthood, and who still had these hearing enhancements. This suggests that the period (or certainly the chief period) for plasticity in this system is in early childhood. For old brains, it's a very different story. But then, millions of older people who are blind or partially-sighted can't hear properly, either.

Pamela, Julie and Colin fall into this category. Exactly how big it is no one knows. Somewhere between one in twenty and one in five people aged over seventy are thought to have significant impairments in both senses; in the UK, the Department of Health estimates that perhaps 1.1 million people are affected in this way.

The interviews with elderly people, most of whom were in their eighties, highlighted all kinds of knock-on effects. Because being able to hear well is so important for communicating with family and friends, many felt that their hearing impairment was more disabling than their vision problems. It certainly made it hard for them to interact with other people:

Jackie (88) hosts regular meetings in her home for church members but sometimes struggles to participate in the group conversation. She asks them to speak up but as soon as they get excited or engrossed by what they're saying they start talking too quiet again with their heads down. She was given an aid that she has to press when she can't hear, but it doesn't help much so is hoping that better hearing aids, when she gets them, will enable her to hear. She said being in a group is difficult because they don't like to speak in that domineering (loud) way – they like to speak confidentially.

Of course, it's difficult to isolate the influence of individual aspects of being old on psychological wellbeing. Sensory deteriorations. Ill health. Bereavement. Muscle and bone weakening. Cognitive impairment . . . Few elderly people have experience of only one of these things. However, a recent analysis of available research concluded that having both hearing and vision loss is, in and of itself, associated with a significant increase in the risk of depression. And there's research finding that sensory decline can drive personality changes, too.

In one large, four-year study in the US, when disease, depression and other variables were all accounted for, declines in hearing and vision over time were related to steep reductions in extraversion, agreeableness, openness and conscientiousness, and less in the way of the normal decline in neuroticism (which typically happens for older people who can still see and hear well).[29] In fact, the researchers concluded, sensory functioning was a better predictor of personality change with age than disease, or levels of symptoms of depression.

By making social interactions difficult, sensory loss could make people less inclined to be sociable – less extraverted – and also more grumpy (less agreeable). Poor hearing and vision could discourage older people from straying from the familiar (less open) and make it harder for them to achieve what to most of us would be minor goals (like leaving the house or cleaning up), potentially reducing conscientiousness. All in all, declines in hearing and vision are, as the researchers put it, associated with 'maladaptive personality trajectories' in older adults.

Reduced agreeableness, openness and extraversion clearly have the potential to weaken a person's existing social ties, and their ability to form new ones. Lowered conscientiousness could have more immediate impacts on health, making it less likely that they will stick to exercise or drug regimens, and more likely that they'll start drinking too much alcohol.

All of this highlights just how profoundly important our senses are to what we do, who we see, how we behave and how we feel. If our senses deteriorate, all of these things deteriorate, too. When it comes to hearing, there's also growing evidence that typical age-related progressive impairments[30] can have profound impacts on something even more fundamental: our memory. In fact, there's growing, and worrying, evidence that this type of hearing loss — which starts with the loss of high-frequency hearing, through to mid-frequency and finally low-frequency — raises the risk of one of the most terrible afflictions associated with older age: dementia.

Back in 2011, a US team published the first work suggesting a link between age-related hearing loss and dementia.[31] This alarming finding prompted all kinds of follow-up work, which substantiated the link and further explored the risk.

In 2018, a team at Ohio State University reported findings from a study that in fact started out in quite another direction. The researchers wanted to investigate how people's brains respond to hearing complex versus simple sentences. Before they could start the study proper, they had to test the participants' hearing, just to make sure that all could hear well enough to take part. These people were young – aged nineteen to forty-one – and, as expected, they all passed the hearing tests. But when the final data analysis was done, something unexpected popped out.

In healthy young people, the left hemisphere of the brain handles language comprehension. For the volunteers in this study with good hearing, this was exactly what the team observed. However, for those with subtle hearing loss (at a level you wouldn't even be aware of), when they processed spoken sentences, there was activity in the right hemisphere, as well. A switch to a dual-hemisphere response to a familiar language doesn't typically happen until a person's fifties. For this group of younger people, 'their brains already know that the perception of sound is not what it used to be and the right side starts compensating for the left,' explains investigator Yune Lee. These findings worry Lee: 'Previous research shows that people with mild hearing loss are twice as likely to have dementia. And those with moderate to severe hearing loss have three to five times the risk. We can't be sure, but we suspect that what happens is you put so much effort into listening, you drain your cognitive resources, and that has a negative effect on your thinking and memory.'[32]

If this is right, and a slight hearing impairment causes a cognitive drain to begin sooner, this might increase the dementia risk – or, at least, the symptoms of dementia might appear earlier than they would have otherwise.

In 2020, a German team exploring changes in the brains of mice published some startling results.[33] Sudden sensory loss is known to trigger a widespread reorganisation of key brain areas. At root, this is a good thing – it's evidence that the brain is changing, in a bid to cope with the challenge.[34] But there can be some temporary ill effects. Information from our senses is critical to our formation of memories. (Think about what you did yesterday evening . . . your memories prominently feature what you saw, heard, tasted, and so on, don't they?) There are strong links between the parts of the brain that process sensory information and the hippocampus, our most important memory structure. The sudden loss of a sense can therefore cause disruptive changes in the hippocampus – which manifest as memory difficulties. Eventually, this disruption settles, limiting the disruption to memory.

A gradual deterioration – as happens with age-related hearing loss – is different, though. The brain has to constantly adapt to an ever-changing sensory input. And, the work on mice revealed, this means that disruptions in the auditory cortex and the hippocampus can't settle, either. 'This is very likely why memory becomes impaired,' says Denise Manahan-Vaughan at Ruhr University Bochum, who was involved in the study.

This process does not itself cause Alzheimer's disease (or vascular dementia, which is also relatively common). But in its acting as such a drain on the brain's resources and interfering with memory, hearing loss could make it much harder for the brain to cope with another challenge, such as the protein plaques and tangles that are associated with Alzheimer's disease. And this has practical implications, the researchers argue: 'I think my study shows that it is extremely important to start wearing a hearing aid when hearing loss becomes apparent,' says Manahan-Vaughan. 'I wouldn't go so far as to say this will prevent dementia – this is another physiological process entirely

– but wearing a hearing aid is likely to slow down the progress of memory impairments that occur in healthy ageing, simply because one reduces the demands on the brain to adapt to the progressive loss of a sensory modality.'

For all kinds of reasons, then, it's vital to protect our hearing. To do that, we have first to understand the threats.

Auditory hair cells are delicate creatures – highly sensitive and vulnerable to damage. However, a sudden very loud noise – even a bomb blast, say – won't necessarily rob of your hearing. If the three tiny bones in the middle ear are spared and the hair cells survive the initial trauma, resulting hearing difficulties can rapidly resolve.[35] But if the bones are damaged or the membranes within the cochlea so violently shaken that hair cells are killed, it's a different story. When a hair cell dies, it is not replaced, and the hair cells near the base of the cochlea, which detect high-frequency sounds, are most vulnerable.

For most people with impaired hearing, their hair cells have been lost gradually, because of long-term exposure to damaging, but less extreme, loud sound. At least, this is the message from major health bodies, including the World Health Organization.[36]

According to WHO guidelines, the threat to hearing is a function of loudness and duration: twenty-eight seconds per day at a loud rock concert at around 115 decibels, or more than fifteen minutes a day using a 100-decibel hairdryer (or a motorbike), for example, and you risk damaging your hearing. The WHO estimates that more than 1.1 billion teenagers and young adults around the world could be at risk of hearing loss because of unsafe levels of such recreational noise.

The US National Institute on Deafness and Other Communication Disorders stresses that long or repeated exposure to sounds at or above eighty-five decibels (which is the cinema range) can cause hearing loss. The louder the noise, it's argued, the shorter the amount of time it takes for damage to occur. The everyday sounds of modern life represent, then, a multifarious threat to our hearing.

This is not a message that I'm going to contradict. But there are some scientists who think it isn't quite right. Gerald Fleischer at the

University of Giessen, Germany, is one. Fleischer has spent many years evaluating the hearing of people from around the world, and he's reached some controversial conclusions.[37]

If I were to ask you: Who's most likely to have poor hearing – a construction worker in Berlin, or a nomadic yak herder, I imagine you'd go for the construction worker. I would have done. But Fleischer has found that both these groups of people have similarly poor hearing. He does not agree with the idea that regular exposure to everyday types of sound, such as hairdryers, is incrementally damaging. He believes that sudden, very loud noises are the chief destroyers of hearing, but also that our ears need some level of regular stimulation to be able to cope with loud noise.

Though we are born able to hear, it takes time for our ability to rapidly distinguish different sounds to develop. 'Young children do not hear well, because their auditory system needs time and exercise to develop properly, and needs a wide range of stimulation,' Fleischer writes. 'Auditory sensitivity increases up to the age of about twenty years. And nomadic people, just sitting on the grass, guarding some sheep or yaks do not hear well, because their way of life is character-ised by auditory deprivation.' When firecrackers are then let off – as he observes happens during occasional celebrations in these villages – the preceding quiet is like adding a knuckle-duster to that sonic fist.

That doesn't mean, however, that lots of background noise is always beneficial. Anyone who's ever collected a child from nursery school knows just how unbelievably loud a group of three-year-olds can be. And a recent survey of almost 5,000 nursery school teachers in Sweden found that 71 per cent had 'auditory fatigue' (meaning, for example, that they just couldn't bear to listen to the radio after work) compared with 32 per cent in a comparison group of women with a range of other jobs. Almost half of the teachers said they had trouble understanding speech, compared with a quarter of the other women. And almost 40 per cent said that at least once a week they experienced discomfort or physical pain in their ears from everyday sounds.[38] None of this means that their hearing had been irreparably damaged (their rates of hearing loss and tinnitus were not much different than for the other women), but it's clearly not good for them.

It's worth adding that modern, everyday kinds of noise exposure threaten more than our sense of hearing. The everyday din of traffic alone causes not only tinnitus but sleep disturbances, heart disease, obesity, diabetes and even cognitive impairment in children, according to another recent WHO report. In Western Europe alone, the impacts just of traffic noise have been equated to the annual loss of at least one million healthy years of life. No wonder the excess of noise that surrounds so many of us has been called a public health crisis.[39]

If you want to protect your hearing, avoiding very loud noises is clearly a stand-out strategy. There's also evidence, though, that some more surprising factors come into play. For example, obese people are more likely to suffer hearing loss,[40] and regular physical activity and a healthy diet are protective.[41] But maximising hearing isn't just about conserving what you have. You can also train your brain to get better at processing sounds.

Have you ever been in a crowded bar, frustrated that you can't hear the conversation? To home in on what someone's saying, especially in a noisy environment, your brain has to precisely process patterns of incoming sound that make up a word – a challenge that we know typically gets harder in older age.

Nina Kraus is an auditory neuroscientist at Northwestern University, Illinois, and an amateur musician with a fondness for 'Smoke on the Water' (at least, she likes to play it at conferences). And she's shown that even short-term training programmes can make a real difference to your hearing ability.

Over eight weeks, a group of healthy fifty-five- to seventy-year-olds spent a total of forty hours engaging in some pretty challenging computer-based listening and memory tasks. They practised, for example, telling apart similar-sounding syllables and words, like 'bo' and 'do', 'big' and 'bid,' or 'muggy' and 'muddy', which came at them ever more rapidly as they improved. They also had to repeat sequences of syllables and words. At the end of the training period, Kraus found that the volunteers in this group were notably better at perceiving speech sounds against background noise than a group that hadn't received the training. The delays in sound processing that come with age had been partially reversed.[42]

When you consider older people as a group, there is a lot of variation in hearing ability, Kraus notes. Despite the apparent risks, older musicians (not necessarily full-time professionals, but people who know how to play an instrument and like to play it) tend to be especially good. In fact, Kraus has found that musical training enhances performance on a range of listening tests. Whether you're looking at brain activity during tasks involving discriminating between frequencies or hearing what someone's saying in a noisy environment, it can be hard to tell apart images from a healthy young person and an old musician. 'The brain's response to sound can be enhanced by making music on a regular basis,' Kraus says. 'Biologically, the response to sound in a musician can look like what you'd expect in a younger person.' For her, the evidence that playing music trains and strengthens the ability to make sense of sound is clear.

Armed with a good sense of hearing, extraordinary things are possible. One of the most remarkable videos that I've ever seen is of an American man called Daniel Kish riding his bike through city traffic. Kish has been blind since childhood – but he can sense when to turn corners and when to brake because a car is pulling out, simply by hearing. A few times each second, Kish taps his tongue sharply against the roof of his mouth, making a clicking sound. By listening closely to the bounce back of those sound waves, he can, bat-like, map his physical environment.

Kish's skill is incredible. In the video, while walking with the presenter, he demonstrates that he can easily hear a passageway between two buildings. As the pair start to pass a series of slender, tall columns, Kish says, 'And then we have stuff, basically. Tall stuff in the way. Posts or something, right in the middle of the walkway.'[43] He reports that he can glean enough information from click echoes even to determine whether a fence that he's walking or cycling past is made of metal or wood.

If you are blind and you're listening to this book, even if you haven't heard of Kish until now, this may strike you as a virtuoso performance of a strategy that you already use yourself. Listening closely to natural sounds or making sounds, perhaps with the tip of a cane or shoes or the mouth, and listening for the echoes, is something that many blind people do – at least according to comments made by a number of blind people in the wake of publicity about Kish.

Whether you are visually impaired or have typical vision, in theory, you could learn to do this type of echolocation, too. Lore Thaler at the department of psychology at the University of Durham has studied Kish and also other expert echolocators, and found that the way they have all independently learned to use clicks is very similar. It's common, for example, for the people she has surveyed to make louder clicks to 'see' behind themselves. Thaler is now teaching these techniques to other people, both sighted and blind, to explore further what happens in the brain as the skill is acquired.[44]

There is some evidence that people who have lost their sight early in life tend to find it easier to learn to echolocate than people who have lost their sight only recently, or who can see. But experiments have also revealed that an individual's ability to focus their attention is also important. This makes sense, of course: at least when you're learning, you have to pay close attention to slight variations in sounds to perceive them.

If you do have typical vision, I have a question for you: now you know that it's perfectly feasible to use sounds that you make to sense your physical environment, do you think perhaps you might already be doing it – only you hadn't realised?

This thought certainly occurred to Tim Birkhead, a behavioural ecologist and renowned ornithologist at the University of Sheffield. In 2012, Birkhead published a wonderful book called *Bird Sense*.[45] In the course of doing research for the book, he investigated the echolocation abilities of some birds, including oilbirds living in Ecuador, which click and shriek while entering or leaving the dark caves in which they roost. While echolocating, a bat uses sounds that are far too high a frequency for you to hear. In contrast, the oilbirds' sonar sound is definitely audible to our human ears. It's a rougher and readier system than that of bats, but it's fit for purpose.

Birkhead was thinking about these oilbirds, and Daniel Kish, and other visually impaired people he'd read about who'd learned to use echolocation to cycle, and he found himself wondering about a particular room on the corridor near his office at the university. It's sparsely furnished, with a tiled floor and a badly-fitting wooden door that scrapes loudly when it's opened, creating a sound that reverberates around inside. To see whether anyone's in there, you have to step on in. Birkhead decided to try to guess whether anyone was inside just by paying close attention to the noise caused by the scraping door. He did this on many occasions. And to his surprise, he found that he was right about 85 per cent of the time. A body inside the relatively small room takes up space, subtly altering the way the scraping sound reverberates. And Birkhead could hear it. 'I was shocked at how accurate I was,' he tells me.

It's an anecdotal report, for sure. But, as Birkhead and I discuss it, we reflect on how knowledge about what our senses are capable of expands our perceptual horizons. Perhaps, in your life, you don't need to use sound echoes to try to work out if someone else is with you in a room. But how wonderful that, in principle, you probably can.

3

Smell

How to sniff out dangerous people — and improve your sex life

Man smells badly.

Aristotle, *De Anima*

Do you have a SCENT PRESERVATION KIT?

K9 Ally hopes that you do.

Last night K9 Ally and her handler, Deputy Justin Williams success-fully tracked a missing endangered elderly woman with dementia. She had been missing from her Sugarmill Woods home for about 2 hours.

Her kit was completed about 2 and a half years ago. The scent kit helped K9 Ally and Deputy Williams locate her in less than 5 minutes!

The woman was returned home safely and K9 Ally was rewarded with a special treat, a tasty vanilla ice cream cone!

Great job Deputy Williams and K9 Ally!

This 2017 Facebook post from the Citrus County Sheriff's Office in Florida, USA, was picked up by media outlets around the world.[1] Of course, we're all familiar with the concept of a sniffer dog tracking a person's scent. But it had been more than two years since this woman had rubbed a piece of sterilised gauze onto her underarm, then sealed it away in a jar. Great job indeed, K9 Ally!

Dogs aren't the only animals that can learn to track a scent across the ground, however. A team at the University of California, Berkeley, recently asked a group of human volunteers to go on their hands and knees in a grassy field. They were blindfolded, deafened with earmuffs, and given thick gloves and kneepads to prevent them from using touch to sense the terrain. Remarkably, most learned to sniff their way along a smell trail, 'zigzagging back and forth across the path like a dog tracking a pheasant'.[2]

A result like this would have been a huge surprise to Aristotle. For him, the human sense of smell verged on the pitiful. Certainly, he felt it was inferior to that of other animals. 'Man perceives none of the smell-objects except the painful and pleasant ones, as his organ is not accurate,' he stated.

Centuries later, the German philosopher Immanuel Kant developed Aristotle's disdain for human smelling into something closer to a rant: 'Which organic sense is the most ungrateful and also seems to be the most dispensable? The sense of smell. It does not pay to cultivate it or refine it at all in order to enjoy; for there are more disgusting objects than pleasant ones (especially in crowded places), and even when we come across something fragrant, the pleasure coming from the sense of smell is always fleeting and transient.'[3]

John McGann at Rutgers University has traced the origins of the idea that we're terrible smellers more directly to the Victorian era – and to Paul Broca, the French neuroanatomist. Broca, who died in 1880, classified humans as 'non-smellers' not as a result of any sensory testing but because, McGann has written, he believed that the evolutionary enlargement of the human frontal lobe gave human beings free will at the expense of the olfactory system. This idea influenced many other scientists of the time, who ran with it. It gave rise to the claim that we have 'microsmaty' (tiny smell), something that Sigmund Freud argued rendered us susceptible to mental illness. 'Even today, many biologists, anthropologists and psychologists persist in the erroneous belief that humans have a poor sense of smell,' McGann goes on.[4] So just how misguided is this idea?

For you to smell something, molecules given off by an object must reach the smell receptors high up in your nasal cavity. These receptors are located on the tips of olfactory neurons, which send their signals straight to the brain.[5] At the moment, no one can look at a molecule and say, based solely on its structure, how it will smell or even if it will smell at all. All we know is that for something to have a smell, its molecules must easily evaporate so they can be carried in air and inhaled – but they must also dissolve in your nasal mucus, where your suite of about 400 different types of receptor tip lurk.

Since we know that chemical detection is an ancient sense, it comes as no surprise that even very simple organisms engage in the same process. For example, the bacterium *Bacillus subtilis*, which is found in soil, as well as your gut, senses airborne ammonia given off by rival bacteria using a molecular 'nose' in its membrane.[6] For you, only one particular type of smell receptor is active at the end of each olfactory neuron. But each will bind to a limited group of similar molecules – and one smelly molecule can bind to more than one receptor. This makes for a complex pattern of stimulation every time you step out into a garden, or shove dirty clothes into the washing machine, or open a tin of tomato soup. Your brain then has to read these smell 'barcodes' and interpret what they mean.

The human olfactory system is generally similar to that of a rat or a dog, say. A dog does have about twice as many functioning types of olfactory receptor as we do – and this is often held up as the reason why dogs seem to be so relatively amazing at smelling. But we have more complex olfactory bulbs (the brain regions that handle initial processing of smell signals, at least in most people*),[7] and a superior orbitofrontal cortex (the region that makes sense of smell signals from your nose, and helps you work out what to do about them – run to turn the oven off, perhaps, or put on a fresh set of clothes).[8] In any case, the sheer number of olfactory receptors that an animal can lay claim to does not necessarily translate into an obviously better sense of smell.

Joel Mainland, a smell scientist at the Monell Chemical Senses Center in Philadelphia, USA, notes that cows have more types of olfactory receptor than dogs – about 1,200 compared with 800 – but it's far from clear that cows are significantly better at smelling. Also, it was thought that humans can detect perhaps only 10,000 different scents, but there has been a radical rethink. In 2014, a paper published in the journal *Science* estimated that we can detect more than a trillion smells.[9] There's some debate among smell scientists about

---

* About 4 per cent of left-handed women (but not men) lack olfactory bulbs. Many of these women can't smell – but about one in eight of them can.

just how accurate this estimate is, but whatever the true figure, it's massively more than anybody thought.

The poor reputation of humans, Mainland argues, might be down to the fact that we spend comparatively little time actively sniffing and so training our sense. Unlike dogs or rats, we don't spend much of our time with our noses down by the relatively scent-soaked ground.

Our genes, however, tell a much bigger smelling story. Looking at your smell-related DNA is almost like surveying the craters on the Moon and realising that, though it may look calm and settled now, it has had a crazily tumultuous past.[10]

For each working smell receptor gene, you have another that, over time, has lost its function. This is thought to be because, on the evolutionary journey from distant ancestor to us, the suite of scents that each species in the chain needed to detect to survive and thrive changed. Scents that really mattered to the synapsid that gave rise to mammals, say, just didn't matter to a species a few steps further on in our lineage.[11] So when mutations happened that stopped receptors for these smelly molecules from functioning, it made no survival difference. We still carry those genes – but as damaged baggage. At the same time, new olfactory receptors evolved. And this combination of loss and gain has left us with our own particular 400 or so functioning types.[12]

Research into which molecules these receptors respond to helps to reveal what smell is 'for'. Some of those functions you'll be at least roughly familiar with. But others are only now being revealed – and their impacts on our physiology and our psychology are nothing short of astounding.

As we all know, smell is important in relation to food. Our sense of smell helps us to decide whether something will be safe and nutritious, or dangerous, or even deadly. In fact, as I'll explore more in the next chapter, most of what we think of as 'flavour' is down to smell, rather than taste. Smelly chemicals given off as we chew or drink drift up from the mouth via the nasopharynx into the nasal cavity. It's our sense of smell that allows us to recognise and enjoy chocolate in a pudding – or to decide that it would be wise to spit out that mouthful of fish.

Fresh fish doesn't smell particularly fishy. A pungent fish smell triggers suspicions, then, that your dinner is not safe to eat. But even this apparently straightforward response to a smelly chemical can have fundamental effects not just on our actions, but our thoughts.

When we're not convinced about a statement or situation, we might say that it 'smells fishy'. English-speakers are not alone in this; apparently, the same metaphor is used in over twenty languages. Norbert Schwarz, a professor of psychology at the University of Southern California, Dornsife, recently led research showing just how far the 'smells fishy' metaphor can go.[13]

Participants in this study were given a document to read, then asked questions about it. While they read the document, they were sitting at a desk that either had – or had not – been spritzed with a little fish oil. Schwarz found that the fish-oil group were more likely to spot deliberate logical conflicts that had been inserted into the text. Whether they were conscious of the scent or not didn't matter. Either way, it seems to have raised their suspicions generally, and as a result they were more critical of the content of the document.

There's other research that supports this finding, in showing that even faint fishy scents make us less trusting of other people, too.[14] As Schwarz comments: 'If I'm distrustful, then I'm thinking, "Something's wrong here." And then I have to think more critically and figure out what is wrong.'

Work like this reveals the deep and sometimes surprising connections between our sensory perceptions and our unconscious reasoning. But the practical message from these particular studies is that if you want to boost critical thinking in your office, you could do worse than spray a little fish oil around. Just make sure you wash your hands before going into a meeting.

Other odours – or, rather, what we often think of as being odours – meanwhile, may help with office productivity. Peppermint, for example, has some stimulant effects. But, as I'll explain in Chapter 9, these effects don't occur via our sense of smell.

As with vision and hearing, there are many genetic differences between people in how we smell. It's even been suggested that if you want to find two people with the same olfactory receptor genes, you'll need to get hold of a pair of identical twins.

In many ways, it's easier to look at gene-related deficits in smelling, rather than surfeits. Lots of specific scent insensitivities have been identified, though each affects relatively few people.[15] Around one or two in 100 people can't smell vanilla, for example, but their ability to detect other odours is not affected.

There are also genetic variations that can make for radically different perceptions of a key target of our sense of smell: other people.

For more than two decades, it's been known that simply by smelling their body odour, we can identify members of the opposite sex who have different immune system genes to our own – and so who would, all things being equal, be a decent choice of mate (because mixing immune system genes would be beneficial for offspring).[16] Work to explore further how body odours might play a role in social relationships has progressed since then, though it remains a small field. Camille Ferdenzi at the French National Centre for Scientific Research, who specialises in the role of body odours in our interactions, has made some important discoveries.

She's found that odours emanating from men's and women's heads smell very different. In her research, men's head odours were more often described using the terms 'fatty, sweat, musk and butter', while women's odour was described as 'more floral, woody, heavy and mineral'. Armpit odours were different again. Male odour was more often described with the adjectives sweat, acid, blackcurrant, green, mineral and spicy. Women's odour was more likely to be matched to these smell terms: earth, floral, fruity, sweet . . . but also faecal and vomit. Ratings for the attractiveness and intensity of head and armpit odours were highly correlated in women, but not in men. This might be because a man's head odour conveys different, perhaps socially relevant, information to their armpit odours. However, Ferdenzi has also found that variations in olfactory receptor genes could easily mean that while I'd find one person's natural scent perfectly pleasant, you'd think it foul.

When it comes to how we smell each other, one particular olfactory receptor has been studied more than any other. It's known as OR7D4, and one of the key molecules it binds to is androstenone. A derivative of testosterone, there's more androstenone in male than female body odour. Leslie Vosshall at the Rockefeller University, New York, has led research that reveals that there are two main human variants of the gene for OR7D4. For those who inherit two copies of one variant, androstenone smells bad. For those with one copy of each, it either smells of nothing, or like vanilla. However, further slight variations in the gene make some people hyper-sensitive to it. To them, it isn't just unpleasant. It's repulsive.[17]

Variations in the OR7D4 receptor also explain individual differences in smelling another molecule that's produced more by men than women, called androstadienone. Androstenone and androstadienone smell similar. Ferdenzi can attest to this, as she's very sensitive to both. For her, the odour profile is a mixture of urine, sweat and musk.

I can't help asking whether she feels this sensitivity influences her perceptions of actual men. She replies by email:

> From a very personal point of view, I have smelled this odour in men only in the uro-genital area, not really from other body parts (not clearly from the armpit, anyway). I do not find it pleasant, but it is like other unpleasant odours from the body (bad breath when we wake up, feet odour . . .), you just have to tolerate it :-)

Ferdenzi shared another strange (again very personal) anecdote:

> I smelled androstenone on my baby's head just after birth (only on my first baby, the second one did not smell like this). It was just during the first 1 or 2 days of life and disappeared then. There is one study showing that there is androstenone in the amniotic fluid.

Exactly what androstenone is doing in amniotic fluid or in the uro-genital area of men, and how it may influence other people, isn't yet clear. But, Ferdenzi adds,

> There is undoubtedly a huge variability between individuals regarding perception of others' body odour, as shown for the major compounds of sweat odour – in terms of levels of sensitivity but also in terms of descriptions or liking. There are many other compounds of body odour that may be important in interpersonal relationships, but we don't know much about them yet.

When it comes to how sensitive we are to a specific odour (rather than the black and white distinction of whether we can smell it or not), there's plenty of evidence of a wide variability.

In one classic study, Andreas Keller, a geneticist at the Rockefeller University, with his colleagues, asked 500 people to rate sixty-six odours for intensity, and also for pleasantness. The results plotted a bell curve. When the team then used very weak doses of these compounds to explore volunteers' physiological responses – to probe their ability to detect these chemicals even when they weren't conscious of smelling anything – the results suggested an even greater variability in how they reacted.[18] People who fall to the extreme-sensitivity end of the curve have been dubbed 'super-smellers'.

This variability in human scent sensitivity – in how much of a smelly compound has to be there for you or a friend to be able to perceive it – isn't thought necessarily to relate to the functioning of the olfactory receptors themselves, however, but to the transmission of signals from these receptors to brain regions that then process them. These variations could be partly driven by genes, but they can also be heavily influenced by smell experience. This experience starts in the womb.[19] By about seventeen weeks' gestation, when you were about the size of an orange, your odour receptors were already mature. From the beginning of the third trimester, you inhaled as well as tasted amniotic fluid. Since the flavours of the mother's diet filter into this fluid, you began to gather hints about what you were likely to go on to eat and drink.[20]

Numerous studies show that these very early smell-and-taste flavour experiences (which are continued if the mother breastfeeds) influence which solids an infant will ultimately savour, or spit out. For example, babies whose mothers drink carrot juice while pregnant and breastfeeding go on to enjoy carrot-flavoured cereal as infants – a distinctive preference that babies previously unexposed to heavy carrot consumption do not share.[21]

Your smell dislikes and likes can, then, be influenced from a very early stage by the foods of your own culture. But there are other important cultural factors that impact on smell. Some societies simply value smell a lot more highly than others. And this can make an enormous difference to an individual's smell life. In theory, we all have the potential to be 'super-smellers'.

In 2003, the year that Anna Gislén published her initial work on the extraordinary underwater vision of the Moken children, Asifa Majid was a young post-doctoral fellow at the Max Planck Institute for Psycholinguistics in the Netherlands. She was becoming ever more interested in the possible effects of culture – and language, in particular – on perception. Vision certainly interested her, but a lot was being done on it. When it came to smell, she realised that virtually nothing was known.

When I meet Majid, she has just started a professorship at the University of York, and she's still unpacking. Already up on the wall, though, is a curious framed montage of side-on photographs of noses. It was a parting gift from the students in her lab in the Netherlands. 'The lab noses!' she explains. It is an apt memento for a researcher who is now internationally known for her work on cultural variation in this particular sense.

Majid grew up in Glasgow, Scotland, the daughter of Punjabi-speaking parents. This led her, she thinks, to develop a keen appreciation of cultural differences, and a desire to understand how they might impact on the way people behave and think.

As a post-doc in the Netherlands, Majid was well aware that Dutch-speakers, like English-speakers, had a relatively low opinion of smell, compared with vision or hearing, and this seemed to be reflected in their daily lives, and their performance on sense tests.

Show the average European adult ten different colours and they'd have no trouble whatsoever in identifying them, Majid knew. Get them to sniff ten different scents, and it would be a very different story. But she had read some anthropological work suggesting that smell was more important in some other cultures. She began to wonder if perhaps smell's lowly status was not universal.

In 2006, Majid started work, with Steve Levinson, co-director of the Max Planck Institute for Psycholinguistics, on an anthropo-logical sense-testing field manual. The pair made up testing kits, complete with colour chips and scents, and gave them to researchers to go off to study people from more than twenty different cultures, each of which had a different language.

Swedish linguist Niclas Burenhult took his kit to a community with whom he already had a relationship, and whose language he knew well. This was the Jahai, a hunter-gatherer society who live in the rainforest near the border between Malaysia and Thailand. Burenhult's intention was to run the full gamut of sensory tests. 'But he came back excited about the smell stuff,' Majid says. 'He had this big database of smell words, and I didn't really believe him in the beginning. I thought he just couldn't be right.'

Keen to find out more, in 2009 she joined him on the first of what would become several field trips. And from the outset, it was obvious to her that smell was far more important to the Jahai than it was in the West. She explains that Jahai babies are named after fragrant things (often, but not always, flowers) and that adults adorn themselves with objects chosen not for how they look, but how they smell: 'So they might have a big, ugly ginger root with tiny little flowers that are nothing to look at, but which smell really fragrant, stuck in their hair, behind their ears . . . A thing's odour properties is what matters.' Many of the Jahai's taboos, too, involve smell. For instance, certain meats should not be cooked on the same fire, and brothers and sisters should not sit too close to each other, because if their smells mix, this will upset the thunder god, who will unleash thunder. 'Over the years, we've been able to establish that smell is relevant for basically everything in their life,' Majid says.

Burenhult and Majid learned that the Jahai language contains about a dozen different words for the *qualities* of a smell. *Haʔét* is the Jahai word for the smell shared by tiger, shrimp paste, sap of a rubber tree, rotten meat, carrion, faeces, the musk gland of deer, wild pig, burnt hair, old sweat and lighter gas. *Ltpit* is the smell of various flowers, perfumes, durian fruit and binturong or bearcat (which smells like popcorn, apparently), among other things; *cŋes* is a smell associated with petrol, smoke, bat droppings and bat caves, some species of millipede, the root of wild ginger, the leaf of gingerwort and the wood of wild mango. *Pʔus* is a musty smell, like old dwellings, mushrooms and stale food (that's one that an English-speaker can identify with; and is the only smell-quality term in Dutch: *muf*); *pl?ɛŋ* is the smell of blood, raw fish and raw meat.

The Jahai show us that it is possible to have an elaborate vocabulary for smell, Majid notes – and that the Western struggle to name odours has to be a result of our culture rather than our biology. In fact, when she and Burenhult presented the Jahai with twelve different smelly compounds, including cinnamon, turpentine, lemon, smoke, banana and soap, they were quickly and consistently able to name the fundamental smell properties. They were far better at describing these odours than an English-speaking comparison group. Even though they had never smelled some of them before, very quickly they were able to home in on distinctive qualities, while the English-speakers were left flailing.[22]

Five years later, in 2018, Majid and Burenhult published follow-up work.[23] They reported that while the Jahai took on average only two seconds to describe an odour, Dutch-speakers took an average of thirteen seconds to find a more concrete reference. Rather than identifying the qualities of an odour, they would have to compare it to something else – for example, describing the scent of a lemon as 'lemony', but in more words.

It's not only northern Europeans who find it hard to describe smells, however. Some other people living in the Malay tropical rainforest struggle, too – and this observation is helping Majid to home in on the most important influences on smell ability.

The Semaq Beri and the Semelai live in the same region as the Jahai. The languages of all three cultures are related, and they live in essentially identical surroundings. The main difference between the Semaq Beri and the Semelai is that the former are hunter-gatherers while the Semelai are settled horticulturists. When given the odour-description tests, the hunter-gatherer Semaq Beri did as well as the Jahai, but the Semalai performed like English-speakers. These results, published in 2018, indicate that it's something to do with being a hunter-gatherer that cultivates smelling performance, rather than language itself.[24]

The fact is, however, that we all have the potential to use our sense of smell to our advantage – and many of us use smell in ways we don't even realise.

*

> [Healthy excrement] shall be reddish and not too smelly . . . Those excrements most indicative of death are black, fatty, livid, watery or ill-smelling . . .Those urines that are most indicative of death are smelly, watery, black and thick. [25]

In the time of Hippocrates, around 400 BC, doctors knew full well that the smell of faeces and urine could provide useful clues to a patient's health (or otherwise). But our bodily waste wasn't the only subject of the keen physician's nose. The particular smell of pus, vomit, ear discharges, fevers and blood clots could also be informative, as could asking a patient to cough and spit on a hot coal, then sniffing the air around it. The idea that we can in theory sniff out sick people goes back millennia.[26] What has changed very recently is our understanding of just what it is about sick people that we can smell – and the impact this has on us.

Today, some metabolic disorders and infections, such as tuberculosis, are known to have very distinctive scents. When someone with TB coughs, they exhale compounds produced by the bacterial pathogen *Mycobacterium tuberculosis*. And if the TB is advanced enough, these compounds can be easily smelled by others.

By altering metabolic processes in cells, various diseases can change the way we smell, too. Dogs can sniff out ovarian cancer in blood samples and prostate cancer in urine samples, for example.[27] And at a 2015 conference in Cambridge on the use of animals to sniff out disease, I first heard rumour of a woman with an extraordinary ability. She was, it was said, able to smell Parkinson's disease, even before symptoms appeared.

This mystery woman turned out to be Joy Milne, a retired nurse living in Perth, Scotland. At a Parkinson's UK session hosted by Dr Tilo Kunath, a researcher at the University of Edinburgh, she mentioned that she'd noticed a 'musky' smell on her husband six years before he was diagnosed with the disease. She'd then gone on to recognise this smell on other patients.[28]

Kunath was intrigued. In collaboration with academics at the University of Manchester, he set out to study Joy Milne's abilities. First, they had to establish whether she really could smell the disease. So the team recruited patients and non-patients to sleep in identical T-shirts. Milne was then given these one at a time to smell. Though she correctly identified all the patients, she also put one of the non-patients in this group. One false positive didn't seem too bad. But, in a development that astounded the researchers, this individual later contacted them with the news of a Parkinson's diagnosis. Any doubts about Milne's ability were laid to rest.

The team went on to investigate what, exactly, she was smelling. And in 2019, they reported their discovery of a Parkinson's 'signature' of particular volatile chemicals found in the sebum, the oil of the skin. This includes compounds that they think indicate altered levels of neurotransmitters, including dopamine, already known to be involved in the disease.[29] Ultimately, they hope this work could lead to a test that can diagnose Parkinson's well before muscle tremors appear.

Although Milne's ability to identify Parkinson's by smell is rare, most of us likely can sniff out illness in other people, even before they're aware that they're sick. There's even evidence that this uncon-scious recognition prompts us, again unconsciously, to do the survival-sensible thing – and steer well clear of them.

Research at the Monell Chemical Senses Center has shown that just hours into an infection, the resulting inflammation alters the body odour of a mouse, and that this can act as an alarm to others.[30] In 2017, a team at the Karolinska Institute in Sweden reported evidence suggesting that something similar may happen in people. The researchers took photographs of the faces of nine healthy women and nine healthy men and also collected body odour samples from them. Then they injected some of the volunteers with a mild toxin, which triggered an immune reaction, to make them 'sick'. (They didn't actually feel sick, but their bodies were mounting an immune response.) Others got an injection of a weak salt solution. A few hours later, these people were again photographed, and samples of their body odour were taken.

A separate group of volunteers were then put into an fMRI brain scanner and shown the 'sick' and 'healthy' photos while smelling either 'sick' or 'healthy' body odours. Each time, they were asked how much they liked the person in the photo.

The team found that 'sick' faces were liked less. This suggests that we can somehow pick up on the cues of the very early stages of infection using vision. But when 'healthy' faces were paired with 'sick' body odours, they were liked less, too. This was despite the fact that the raters couldn't consciously perceive a difference between the body odour collected while volunteers were 'healthy' and while they were 'sick'. The brain imaging backed up their liking responses: when the raters smelled 'sick' versus 'healthy' body odour, patterns of activity in their brains were different. In other words, they could unconsciously sniff even this slight difference, too.[31]

Work at Monell, published in 2018, went one step further – at least for mice. It revealed that mice exposed to others that had been injected with a similar toxin to the one used in the human study took on the smell of the 'sick' mice.[32] This suggests that the physiological response of the injected mice to the toxin was being replicated in their cage mates. Smell seems, then, to provide a kind of early warning of possible infection, and by gearing you up to be ready for it, you could be better able to withstand it.

Might changes in body odour also signal when an animal – including a person – is about to die?

In 2007, a cat adopted by the Steere House Nursing and Rehabilitation Center in Providence, Rhode Island, made US headlines as a 'furry grim reaper'. An article in the *New England Journal of Medicine* by geriatrician David Dosa had described how Oscar would wander the advanced dementia unit, sniffing patients.[33] Occasionally, he would curl up on a patient's bed. This was an ominous sign . . . almost invariably, that patient would die within the next few hours. In fact, Oscar was so unerring that whenever he picked a person to snooze with, staff would call the family in.

It would be fascinating, of course, to know what Oscar was smelling. It would also be interesting to know whether anything like the sick-smell transfer observed in caged mice happens to nurses and doctors. Do they, too, take on the odour of patients with high levels of inflammation? If so, does this physiological response help to protect them? And can it be unconsciously smelled on them by other people? Future research will tell.

Detecting the chemical giveaways of something good to eat, the presence of others like us, potential danger – these are all functions of smell with ancient origins. Before I move into perhaps more surprising ways in which we use olfactory receptors, it's worth bearing in mind that these receptors evolved to detect chemicals of interest. And chemicals of interest aren't necessarily always outside our bodies.

In 1992, a paper published in the esteemed journal *Nature* reported an extraordinary discovery. Human odour receptors, which until then had only ever been recorded in the nose, had been spotted elsewhere: in the tissue that gives rise to sperm.[34] Further research went on to reveal the presence of various of our odour receptors in sperm themselves.[35]

This immediately prompted the question: Why – what are they doing there?

It seems that, among other things, they allow a sperm cell to sniff out and so follow the chemical trail of an egg. As the marine biologist Donner Babcock at the University of Washington has noted, the sperm of sea urchins, and other sea-dwelling invertebrates, are attracted to chemicals produced by the target eggs. This external chemical egg hunt seems to have an internal equivalent in humans, and no doubt in other animals, too.

For a long time, this smelling-sperm discovery remained almost an oddity. But about fifteen years on, it gave physiologist Jen Pluznick some reassurance that what she had just observed in her lab wasn't crazy.

Pluznick was then a post-doc at Yale University. She was in the initial phase of research into polycystic kidney disease, a leading cause of kidney failure. While looking at the activity of genes in healthy and diseased kidney cells taken from mice, she was stunned to realise that some of the active genes were for known olfactory receptors. 'At first, I thought that didn't make a lot of sense – because olfactory receptors should be in the nose, right? But my advisor, who was wiser than me, sort of looked at me and said, "But that would be really cool, right?" And I was, like, "Yeah, that would be *really* cool . . ."

Pluznick immediately switched her research focus to studying olfactory receptors in the kidney. To date, she's found ten of them.[36] One, called OR78 in mice (and, a little confusingly, OR51E2 in people), seems to play an integral role in the regulation of blood pressure.

Your kidneys filter all your blood about thirty times a day, removing toxins into the urine, and reabsorbing what you want to keep – such as glucose, and some water and salts. They also help to manage your blood pressure, by regulating the volume of your blood. When that pressure is high, the kidneys get rid of more salt and water from the blood, reducing its volume and lowering the pressure. The reverse happens when blood pressure is too low.

Pluznick has found that OR78 is present not only in the kidney but in blood vessels too. It binds to short-chain fatty acids, which are released when bacteria in the gut digest starch and cellulose from plant-based foods. 'We think it causes vasodilation,' she explains. 'I think around the gut that makes sense. If you just ate a meal, and it's being digested, you want to increase blood flow to the intestines, to make sure you absorb all the nutrients . . . On a local level, that makes sense to me.'

Exactly how it works in the kidney as part of body-wide blood pressure management isn't yet clear, though.

Pluznick and her colleagues have also found that another odour receptor present in kidneys influences the action of a protein that regulates how much glucose is reabsorbed into blood. As it happens, this protein is already a drug target for the treatment of type 2 diabetes. (Allowing glucose to escape into the urine is a good thing if levels in your blood are too high. A better understanding of the role of the odour receptor in its activity may lead to the discovery of more effective diabetes drugs.)

Pluznick's discoveries have certainly stoked interest in the role of smell receptors outside the nose. A number of different labs are now explicitly investigating this – a big change, compared with when her first work on this subject was published in 2009. And these other teams, too, have made some arresting discoveries.

So-called 'extra-nasal' odour receptors have now been found in a whole range of tissues, including the tongue, skin, lung, placenta, liver, heart, brain, kidney and gut.[37] Exactly what they're doing in all these places isn't yet known. In the brain, there's some evidence that they play a role in the response to injury. Several neurodegenerative diseases, including Parkinson's disease, are associated with abnormal expression of olfactory genes (some patients lose their sense of smell well before they develop movement problems).[38] Could problems with olfactory receptors in the brain be playing a role in the progression of the disease? No one knows. But it's a question that's being explored.

In your gut, odour receptors that are activated by compounds found in various food spices have also been identified. They seem to play a role in 'gut motility' – the movement of food through the intestinal tract.

'I think the idea that olfactory receptors are found outside the nose is still a surprising thing to most people,' Pluznick says. 'But as more labs are working on it, and more examples get into the literature, my perception is that it's becoming more mainstream.'

If you think of nose-smell as just one tentacle of a deep-bodied drive to detect chemicals of interest, it makes perfect sense, however. 'Smell' might not be quite what we thought it was. But re-name sniffing 'nasal chemical-sensing', and a rose will smell just as sweet.

As is the case with hearing, our sense of smell is extremely vulnerable. Anything that stops volatile chemicals from getting up to your smell receptors, or damages those receptors, or interferes with the olfactory neurons' ability to get signals to the brain, will impede your ability to smell. At its worst, this can mean a total absence of smelling, known as anosmia. And for the estimated one in thirty-three people who have never been able to smell or who have lost that sense, the impacts on their lives can be as profound as they are unexpected.[39]

Nick Johnson skims the lunch menu at the White Dog Café, a warren of little rooms and ante-rooms in Philadelphia's university district. He orders the tacos and we get a beer that's on tap. It's called Nugget Nectar, and it's produced by the local craft brewery that Nick's worked at for the past ten years. Nugget Nectar used to be his favourite beer. 'It has a real nice balance of sweetness and hops. But now,' he says, and his face falls, 'it's a shell of its former self to me.' He can describe what it smells like: 'piney', 'citrusy', 'grapefruity'. But he can't smell it any more.

Nick, who's thirty-nine, can pinpoint the moment he lost his sense of smell: 9 January 2014. He was playing ice hockey with friends on the frozen pond at his parents' place in Collegeville, Pennsylvania. 'My feet went out from under me. I hit the back right side of my head. I was out.' He suffered a fractured skull, and there was bleeding on his brain.

Given his injuries, he made a remarkably rapid recovery. Six weeks later, he was back at work and, before long, he found himself in a meeting about a new beer: 'We were tasting it, and the others were saying, "Can you smell the hops in the beer?". . . and I couldn't. Then I tasted it. There were guys saying, "It's got this pale biscuity flavour". . . and I couldn't taste it. Then I went and tried one of the hoppier ones . . . and I couldn't smell it. That's when it clicked.'

The stress of the injury and all the medication perhaps explain why Nick didn't realise he had lost his sense of smell sooner. It came as a shock, he says. The trauma of his head hitting the ice was to blame. As olfactory neurons leave the nose, they pass through tiny holes in a plate of bone. A whack to the head can bash them up against the edges of these holes, injuring or even severing them. In the wake of an accident, there's no way of assessing the level of this type of damage, so no way of knowing whether someone who's lost smell will get it back. Nick was told his chances of it returning were somewhere between 5 and 40 per cent.

It didn't take long for Nick to become acutely aware of the impact this was having on his life. A lack of enjoyment of food and drink is a common complaint for people who lose their sense of smell, and Nick, who was a keen cook who loved to entertain friends, certainly experienced this. But for him, as for many other people who develop anosmia, there was another category of loss altogether.

At the time of his accident, Nick's wife was eight months pregnant with their second child. He says that when his daughter was a baby, he'd joke that an upside of his accident was that he couldn't smell her dirty nappies. But what really hit him was that he couldn't smell *her*. We first talked about a year after his accident, and he told me: 'She was up at 4 o'clock this morning. I was holding her, we were lying in bed. I know what my son smelled like as a little baby, as a young kid. Sometimes not so good, but he still had that great little kid smell to him. With her, I've never experienced that.'

That primordial connection to his children, and to his wife, was gone.

Nick found himself confronted by the sudden loss of smell. But, either instantly, or gradually, we all suffer deterioration. We can grow new olfactory neurons – in fact, the reason we don't all lose our sense of smell at a young age is because of the regenerative capacity of the olfactory system. If you could take a close look right now at your patches of smell-sensitive tissue high up in your nose, you'd be likely to find a mixture of mature and immature olfactory neurons, and stem cells differentiating into olfactory neurons or supporting cells – but that patch will not look the same as it did ten years ago, or anything like the way it did when you were born. 'In newborns, there is this really nice clean patch of neural tissue,' says Beverly Cowart, a smell specialist at Monell. 'This gets spotty even by your twenties.'

Breathing in all kinds of damaging chemicals, including air pollutants, can out-pace your ability to self-repair. Areas that used to support neural cells are replaced by respiratory tissue. As you get older, not only do these smell-free zones get bigger, but your capacity to grow new olfactory neurons declines.

'Older' doesn't necessarily mean very old at all. An estimated one in ten American adults aged over forty has problems with smelling. But every single study of smell has found deterioration with age.[40] There is some evidence that sensitivity to certain odours is affected more than others. For example, one recent study found that although people in their seventies were about three times less sensitive to an 'oniony' scent than people in their twenties, when it came to some other scents – including one that was 'mushroomy' – there was no difference between the groups. Another finding from this same study may also help to explain the stereotypical association between elderly ladies and rose scent . . . Compared with the younger volunteers, the older adults needed a concentration of 2-phenylethanol, the principal floral scent compound found in roses, to be *179 times* stronger for them to be able to smell it.[41]

Exactly why some odour sensitivities should be relatively more preserved, at least for a while, while others vanish isn't totally clear. But, though there may be a few exceptions, by your seventies or eighties, your sense of smell will have dropped pretty much across the board, notes Cowart.

To some extent, this is preventable, or can at least be mitigated. Modern life – particularly air pollution, along with certain respiratory viruses (including COVID-19, of course) – poses threats to our sense of smell that are hard to avoid. But so, too, do some of our favourite pastimes. The link between major head trauma and anosmia has long been known. But very recently, it's become apparent that even a minor concussion – the sort you might get from tumbling on the ski slopes, falling off a bike even with a helmet on, or having a minor car accident – can impair smelling.[42]

Our sanitised modern environments are a threat, too. The experience of the Jahai shows just how important it is to engage regularly with a wide variety of odours. How can you do that when you're going from a clean house to a valeted car to an air-conditioned office?

For an insight into the richness of smell life in the past, there's surely none more vivid than this description of an elderly woman, called Thais, by the Roman writer Martial:

> Thais smells worse than the old urine jar of a greedy fuller, just now broken in the middle of the street, or a goat fresh from mating, or a lion's jaws, or a hide from the tanners across from the Tiber torn from a dog, or a chicken that is rotting inside an aborted egg . . .[43]

As Mark Bradley, editor of the wonderful book *Smell and the Ancient Senses*, notes, 'Here Martial evokes a range of foul smells that would be familiar to the inhabitants of early Imperial Rome.'

It's not so easy these days to become familiar with the scent of a goat fresh from mating . . . (And if you're wondering about the 'greedy fuller', fulling was a way of cleansing woollen cloth by working it in stale human urine.) But perhaps you even avoid pleasant odours because you have culturally inherited, whether you're aware of it or not, a 'moralistic denunciation' of fragrance. This can be traced, argues the French historian Alain Corbin, to the ideas of the puritanical, prudish bourgeoisie of the late eighteenth century. [44] For this group, all perfumes symbolised waste and extravagance, while the headier scents were completely unacceptable because they reeked of blatant sexuality.

The idea that perfumes are decadent is far older, however. In his *Natural History*, the Roman writer Pliny the Elder rants about the use of fragrance. He even cites the case of a Lucius Plotius, a man on the run from a sentence of execution, whose perfume gave away his hiding place, adding, 'Who would not admit that such people deserve to die!' [45]

Kate Fox, director of the Social Issues Research Centre in Oxford, has written about Corbin's ideas and more broadly about changing attitudes to scent. In a recent essay, she also makes this thought-provoking point:

> It is interesting to note that the current trend away from heavy, musky perfumes and towards lighter, more delicate fragrances is also associated with a moralistic tendency – exemplified by the rise of 'political correctness', obsession with 'healthy' eating and exercise, the so-called 'new temperance' movement and other puritanical elements. [46]

The fact is that for most of us, even if we occasionally use fragrance or enthuse about the scents of a meal, our smell lives are exceptionally bland. And if we do find ourselves in a stinky situation, we're liable to 'turn our noses up' at the 'stench', whether it genuinely signals a threat to our health or not.

Given what we know about the role of smell in our life, losing it of course brings dangers.

Nick Johnson has first-hand experience of an obvious, if uncommon, one: 'I came home late one night and went into the kitchen and up to bed. The next morning, my wife woke up at seven. She came in, saying "You left the oven on all night! What were you cooking in the middle of the night?"'

Nick hadn't been cooking anything. It turned out that the safety valve on the gas oven had broken. Gas had been pouring into the kitchen for twelve hours straight. When Nick got home, at 1am, the kitchen would have stunk, but he'd been completely oblivious.

This was a case of an immediate risk to life. But a poor sense of smell is associated with a 50 per cent increased risk of dying over the next ten years – something that can't be accounted for by situations like this – and no one is quite sure why. One thirteen-year study of more than 2,000 Americans aged seventy-one to eighty-two found that people who were healthier at the start of the study, but who did poorly on smell tests, had the most heightened risk. The researchers suspect that poor smell is an early sign of other, insidious deteriorations in health.[47]

So avoiding head injury and air pollution aside, what exactly can we do to protect or enhance our sense of smell?

When I first spoke to Nick, he couldn't smell any individual odours. But he did tell me that he'd recently started to detect *something* when an odour was particularly pungent. He could register the presence of something that would have smelled strongly – just not what it was. This suggested to Beverly Cowart that some of his olfactory neurons were still functional and giving a level of feedback to his brain.

Monell's clinic for anosmics, which was set up in the 1980s, had many research aims: to develop tests for anosmia; to try to determine exactly what proportion of the losses patients were reporting were actually in smell rather than taste; to explore the different causes (trauma, viral infection, polyp growth, and so on).

Naturally, Cowart and her colleagues also wanted to learn about rates of recovery. But the only treatment they felt able to recommend to people like Nick – the treatment that's still recommended today – was to regularly sniff a range of smelly substances at different intensities. 'I don't think it really matters too much what you use,' Cowart says. 'As long as you're stimulating the system, it's either going to broadly respond – or it's not.'

For the majority of the clinic's patients for whom there was no simple surgical fix, there wasn't much they could do to help. For this reason, the clinic ultimately shut down.

Some radical new treatments for smell loss are in development, however. At Monell, for example, researchers are working on using stem cells to grow replacement olfactory neurons. Other teams are taking inspiration from the cochlear implant. This device converts sound waves into electrical signals, which directly stimulate the auditory nerve, allowing people who have lost their hearing to hear again. In 2018, a team at the Massachusetts Eye and Ear Hospital directly stimulated olfactory nerves in a group of healthy volunteers; they reported various smell sensations, including the scent of onions, antiseptic, and sour and fruity aromas.[48] It's extremely early days for this research. As Eric Holbrook, the rhinologist who led the study, notes, however, there's currently very little that can be done for people with anosmia. And he wants to try to change that.

Nick Johnson bore in mind advice to keep sniffing different substances. When we first met, he talked about how he was going about this. He didn't have a set routine, but if he was near a strong-smelling substance – if there was lemon zest in the kitchen, for example – he'd try smelling it. He said it was hard, trying to smell but getting nowhere.

Given what he'd said, and the overall poor odds of a recovery, when I catch up with Nick over the phone four and a half years since our last conversation, I'm prepared to hear that his sense of smell hasn't improved.

'Actually, I'm pretty much back to smelling normally!' he says happily. 'For the most part, everything's come back!' I'm so surprised, and glad that I blurt out congratulations. It's the Tuesday after the US Memorial Day long weekend – a big weekend for anyone involved in the beer industry, and, for Nick, before his accident, for cooking for family and friends. When we met, he wasn't really cooking much, but now he tells me: 'Yesterday, I was barbecuing and smoking food and I could smell it as clear as day.' The pleasure in his voice almost pours down the phone.

The strong scents that Nick had tried, so often unsuccessfully, to smell, like the lemon zest, were the first to return, he says. It certainly didn't happen overnight. But now he's almost back to normal at work, taking part in beer tastings just about as well as anyone. At home, too, he's noticed big improvements. Two months ago, he tells me, his wife gave birth to their third child, a second son. Now he can smell his daughter, who was born very soon after the accident, his older son, again, and the baby, too. 'That connection is right back. And with this little guy, I can smell everything – good and bad.'

Nick's job means that he's surrounded by odours, and conversation about odours. Given that sniffing things is the only proven way to improve smell in cases like his, it seems possible, if not probable, that this environment played a role in his recovery. But for others who have lost smell through an accident or illness, his story does show that, in some cases, a blanket loss need not be permanent. And for the rest of us, research does suggest that whatever smell capability you have right now, you have the potential to get a lot better.

In fact, simply by paying more attention to scents, and spending more time with them, you may even be able to develop a perfumer's rarefied nose.

In an untidy back room of an event organised at the Tate Modern gallery in London in conjunction with sense scientists and philosophers at the University of London, I met Nadjib Achaibou, a young perfumer with an infectious enthusiasm for scents.

Achaibou was born in France to Algerian parents, and grew up in Mexico. He now works in London for a company that develops fragrances for everything from perfumes to laundry detergents. He also likes to take on smell side-projects. In one, for Greenpeace, he re-created the scents of the Amazonian rainforest. (Before going to the Amazon, he explains, 'I expected to use "green notes", which smell of grass, of stems, things like that. But there were smells of rotten meat, rotten food, dead corpses! Every step in the jungle is a different smell.')

When he was a child, though his mother was renowned among friends for being a wonderful cook, Achaibou was not especially interested in scents and neither were any of his relatives. In fact, 'in my family, we have a lot of nose problems. I have uncles who cannot smell at all. Smell was not my forte.' He smiles at me. 'Your nose may be better than mine!' For him, it wasn't a talent but a profoundly affecting experience that drew him into his career. Achaibou can trace the igniting of his obsession back to when he was sixteen years old – and to one person. 'I had a girlfriend, and her perfume drove me completely crazy. In a narcotic way. I was like: I'm going to do *that*. And I followed this intuition, this passion.'

After a degree in chemistry, Achaibou took his masters in perfumery at the renowned ISIPCA post-grad school in Versailles. He soon learned a new way to talk about scents: 'When we are young, we are not trained to describe smells in certain ways, so we use words of food or colours, even music sometimes, of textures to describe them. As a perfumer, I had to learn how to put names on smell. That's a lot of what we do in the perfumer school.'

His nose, he emphasises, is trained, not born. And he insists that, for someone who doesn't have the time to learn to identify thousands of scents, as he had to, the best way to enhance your sense of smell is to use it, and to explore it. 'You might say, "Oh, I like pepper." *Why*? Why do I like it? What is it adding to your dish? That's the first step to enhance your sense of smell. If you see a rose, stop and smell it. If you have a friend wearing perfume, smell the perfume and describe it. When you buy a shower gel, a toilet detergent or a perfume, ask questions. Read the marketing materials, but also trust yourself. You might think, yes they say there is rose in that but what I can smell is lemon. But what kind of lemon? Have you tried a bergamot?'

This turns out not to be a rhetorical question. I shake my head. 'Go and taste one! Try to find one. It is a citrus that is from Calabria, one of the southern islands of Italy. It's a very, very localised production of a very specific citrus that shares the same molecules as lavender. It's a citrus note with some floral aspects on top.'

Practise smelling and, he promises, your perceptions of your world will change. It can add intense pleasure to everyday experiences, such as a walk. 'If I see a woman from the Middle East with her black veils, I always go behind her because she is going to smell great. I don't stalk them! But if I pass them, I just smell because they have a perfume culture that we don't have at all. They use a lot of smells. With them it's a way to communicate things they cannot communicate otherwise. They cannot show flesh. But they communicate through scent in a way that is very erotic.'

A more nuanced sense of smell could enhance your understanding of your world, but even the bad smells are important smells, he says: 'A life without stinky smells is like a face without wrinkles, and a face without wrinkles is a face without life!'

It may even improve your sex life. Men born without the ability to smell are known to have fewer sexual partners. And in 2018, a German team reported that people who are more sensitive to smells enjoy sex more.

The researchers first gave healthy young volunteers a standard smell-sensitivity test. Then they asked them about their sex lives. There was no link between the volunteers' smell sensitivity and how often they'd had sex in the past month or how long each of those sexual encounters had lasted. However, people with a keener sense of smell reported finding their sexual activities 'more pleasant', and women with a generally greater sensitivity to odours had more orgasms during sex. 'The perception of body odours such as vaginal fluids, sperm and sweat seems to enrich the sexual experience' by increasing sexual arousal, the researchers wrote in their journal paper.[49]

I ask Achaibou about his girlfriend's perfume – the one that got its fragrant hooks into his brain back when he was sixteen – does he remember what it was? He laughs, as though it would be impossible for him to forget, and replies: 'Dior's Addict.'

My own, British, culture may not have any embedded methods for enhancing smelling (at least, I can't think of any). But others do. In Japan, for example, there is kōdō, the 'way of the incense', a ceremony that involves inhaling the scent of aromatic woodchips, as well as sometimes spices and other plants, which are heated on a mica plate above a small censer bowl.[50]

Around the start of the seventeenth century, games involving this ceremony started to become popular, and many still are today. In one, guests sniff several different aromatic woods, then the host mixes them up, and the guests compete to identify them correctly. Kōdō isn't just about identifying scents, though. Some scents are tied to places, and to smell them 'properly', you must imagine yourself transported to that distant location. Though this isn't an explicit part of the game, pairing scents to places adds a contextual layer. It can help the brain to categorise what could otherwise be one transient scent among a fragrant blur.

For Asifa Majid, practising smelling different odours is important, but so too is the use of words to consciously name and identify these odours. Smelling something and simultaneously thinking about what it is reinforces the link between the two. And having a sizeable smell-term databank may have unexpected implications.

In a recent paper, Majid and colleague Stephen Levinson point to the case of a council flat in Bristol, where a man lived for years with the corpse of a former friend under his sofa.[51] After neighbours complained about a terrible smell, a council warden came to inspect the flat. She put the stench down to an overflowing toilet. Had she been Samoan, Majid and Levinson note, the body almost certainly would have been found: 'Putrescent flesh emits distinctive gases, e.g. cadaverine and putrescine, rather different than the methane of the latrine, but reporting them under a general label "stink" may literally dull our senses – impossible in a language like Samoan that makes precisely this distinction.'

Smell, then, is a sense that we can let languish if we wish. But, as we've discovered, we humans have the capacity to do extraordinary things with our noses. We might not be up to the feats of K9 Ally (at least, not unless we take to moving around on all fours . . .). But by consciously making smell a bigger part of our lives, we can develop a sense that does extraordinary things for us, influencing not only what we eat and enjoy, but also our health, and how we relate to others . . . 'Man smells badly' might have been Aristotle's biggest sense error of all.

4

Taste

It goes way beyond your mouth

> There is the sweet and the bitter, and adjoining the one the
> greasy and the other the salty, and between these there are sour,
> rough, acrid and sharp.
>
> Aristotle, De Anima

Robert Margolskee, a director of the Monell Chemical Senses Center
and one of the US's leading taste scientists, offers me a bowl of
variously-coloured jelly beans. 'Close your eyes,' he says, 'and pinch
the top of your nose shut.' Then he gives me a bean on my
outstretched palm, and asks me to eat it and tell him the flavour.
I'm at a loss. He tries again with another bean. I have to guess
wildly. Banana? (It was not.)

You can easily try this yourself – and you'll discover just how
difficult it is. As we know from the previous chapter, what we think
of as 'taste' is often 'flavour', which is mostly to do with smell.[1]
Vision, too, preps our brains for what to expect. The only genuine
taste-stimulating compounds in a jelly bean are the sugars, and
perhaps the fats.

Though we can smell a vast array of different odours, our taste
perceptions are much more limited. But that doesn't make taste any
less important. It acts, in fact, as our chief nutritional gatekeeper.[2]
It tells us when we should spit something out – because eating it
will probably harm us – or swallow it down, because we need it to
stay alive.*

---

* One taste or many? When it comes to the *perception*, there's a unity, implying
that this should be considered a single sense. However, when you start to look at

Like smell, it is a chemical sense. The main difference between the two is that smell perceptions are triggered by molecules carried to us on the air, while taste receptors handle chemicals that reach us via our mouths.

Well, that's one of their jobs. Because, as with smell, our understanding of what our taste receptors do for us is going through its own revolution. It turns out that your body is peppered with these receptors, and they guard you in ways you wouldn't believe.

The major classes of chemical nutrients that we really need, from an evolutionary perspective, to survive and thrive, are all detectable by taste receptors in our mouths. So too are the most common food-based toxins. Glance back at Aristotle's categorisation of the taste perceptions that this system is capable of generating, and his outline wasn't bad at all. In fact, on this, he was ahead not merely of his own time but even of twentieth-century ideas.

Qualifying as a primary taste is very like getting into an exclusive club. Strict evidence of your suitability is required. Firstly, there has to be a distinct perception – such as 'sweetness' – and secondly, receptors whose stimulation will trigger that perception must have been identified. As of now, only five basic taste qualities have gained admittance to this club and are widely recognised by taste scientists. They are: sweet, salty, umami, sour and bitter (we'll take a closer look at them shortly).

There is also evidence that we can taste starchy carbohydrates (not just the sugar molecules they are broken down into). But the second criteria – the evidence of specific carbohydrate receptors – has not yet been met. There are also plenty of advocates of 'calcium' as a primary taste quality, but again, it's still a taste hopeful. So too are 'watery' and 'fatty' (Aristotle's greasy).

Some researchers argue that there is compelling evidence in favour of formally recognising fat as a primary taste quality. Recent lab

---

where our taste receptors are found and what they do for us, the argument that it should be considered as five independent senses becomes much stronger. In this chapter, for simplicity's sake, I generally refer to taste as 'a sense', while also explaining why it deserves a more nuanced understanding.

tests have demonstrated what Aristotle believed – that we can detect fats, as various fatty acids, via our mouths – and a couple of possible dedicated receptors have been identified in human taste buds. 'Fatty' isn't quite there yet. But of all the current outsiders, it's closest to joining the exalted 'primary taste' ranks.[3]

So to stick with what's widely accepted, receptors associated with each of those five exalted taste qualities are distributed throughout your tongue in groups of fifty to 100, making around 10,000 taste buds. These buds are mostly clustered in little structures called papillae. If you peer at your tongue in the mirror, those little bumps you can see are the papillae.[4]

When compounds from a burger, or a milkshake, say, dissolve in your saliva, they wash over exposed receptor cells. If a cell recognises its target chemical stimulus, it fires off signals to the brain. Taste information goes to the insular cortex (which is involved in emotion). It also goes to other regions, feeding, along with smell information, into perceptions of flavour, and helping to guide learning about what foods to seek out (burgers! milkshakes!) and which to avoid (that *disgusting* lettuce – if you were to ask my kids . . .).

However, we don't only have taste buds on our tongues. They're also in other parts of the oral cavity, including the epiglottis, and the throat. (Anyone who's ever got an aspirin stuck in their throat before being able to swallow it properly will certainly have experienced the throat's capacity for triggering bitter-taste perceptions.)[5]

For a type of sensing that's been recognised for millennia, our understanding of the receptors that underpin taste is surprisingly patchy. Sweet-tasting has probably been best studied. It's now clear that a range of sugars (including fructose in fruit, sucrose in chocolate and lactose in milk), which are sources of easy energy, and so stamped 'desirable' by your brain, activate the 'T1R2/T1R3' 'sweet'-taste receptor. As its name suggests, this receptor has two constituent units, but both must be activated for the full sweet experience.[6]

The compounds responsible for our perception of 'umami' (the Japanese word for 'delicious essence') are amino acids, and in particular an important amino acid called glutamate. Glutamate is found in various protein-containing foods, including cured meats, shellfish, miso and breast milk, and we need it to build cells. Though glutamate was isolated as a trigger of umami by the Japanese chemist Kikunae Ikeda more than a century ago (using the Japanese soup base, dashi),[7] it was only broadly recognised by international taste researchers at a conference held in 1985. It is, then, the newbie in the club. Several different receptors have been identified as playing a role in umami perceptions.[8] One, known as the T1R1/T1R3 receptor (yes, it's half-identical to the sweet-taste receptor), which responds to glutamate, is the best understood.

A 'salty' taste is usually down to the detection of sodium salts in sodium chloride (table salt). Sodium – at the right level – is critical for healthy physiological functioning. Unsurprisingly, then, we tend to find very salty foods unpalatable, but like low to moderate concentrations. The best-studied salt receptor is an ion channel that opens almost exclusively in the presence of sodium salts. However, in 2016, a team that included Margolskee published details of a second salty-taste pathway (in mice). This pathway involves a subset of what are normally thought of as 'sour'-taste cells. It responds to the negatively charged ions in a salt – the Cl⁻, for instance, in sodium choride.[9]

Sour tastes come from acids. They can indicate potentially dangerous bacterial spoilage, which is probably the main reason we have evolved not to like very sour foods. However, some sourness is often appealing – and some fruits, including oranges, green apples and grapefruits, which can certainly taste sour, are also rich in Vitamin C, a compound that we can't synthesize ourselves and need to consume. In terms of the molecular biology, while there are ideas about how molecules that generate a sour taste are detected, this is the least well-understood of the five.[10]

Bitter-tasting is better understood – and also more complicated. Twenty-five different receptors, all members of the T2R bitter-taste family, make several hundred different compounds taste bitter. Some of our T2R receptors are tuned to respond to very specific compounds, whereas others respond broadly to many.[11] A bitter taste can indicate that a vegetable or a fruit contains toxins. All kinds of animals, even oysters – even protozoans, for that matter – will reject bitter foods.[12] Detecting these particular tell-tale chemicals is that basic, and that critical for survival.

However, for us humans, not all bitter tastes should be rejected. If these toxins are present at relatively low levels, and we don't eat an excess of them, the nutritional positives can outweigh the negatives. If you're a picky eater, some examples of bitter vegetables might spring readily to mind: broccoli, kale, watercress, bok choy, kohlrabi and turnip all fall into the bitter-but-good-in-moderation category. They contain plenty of desirable nutrients. Unfortunately, they also have a dark side: a class of compounds called glucosinolates, which are essentially poisons, because they inhibit the thyroid gland's ability to take up iodine, which it needs to synthesize essential hormones. Our 'TAS2R38' bitter-taste receptor recognises glucosinolates (among other compounds). These vegetables can, then, taste off-putting, while being good for us at the level that most of us consume them. (A glass of raw kale juice for breakfast is not recommended.)

Many medicinal plants also taste bitter. Wormwood (*Artemisia absinthium*), which is referred to in medical papyri from Ancient Egypt, and which Hippocrates used to treat menstrual pain and rheumatism, is a classic example.[13] So too is quinine, an effective anti-malarial. Originally sourced from the bark of the cinchona tree, quinine gives tonic water its bitter taste – a problem that British officers in India found could be effectively overcome with a little sugar, lime and gin.

If you consider your own bitter-taste system, it probably evolved to help you to identify medicinal compounds, as well as poisons. However, it's very unlikely that we can taste the difference between 'good' bitter compounds and 'bad' ones, Margolskee thinks. More likely, we have learned which bitter plants make us feel better, and passed that knowledge on via culture.

So far, I've been talking about taste as it's conventionally understood – on the tongue. But now the conventional taste story takes a twist, right down inside your digestive tract.

Recent research has shown that dotted all the way through your oesophagus, stomach and intestines are two different types of 'taste-like' cells. 'Taste-like' is the term preferred by Margolskee, because these cells aren't clustered in buds, and they don't directly lead to conscious taste perceptions, unlike taste cells on the tongue.

However, they do have some of the exact same taste receptors that can be found in your mouth. Understanding how these 'taste-like' cells function can even give us an insight into how to improve our diet.

In the mouth, we use the 'T1R2/T1R3' sweet receptor to detect sugars. In the intestines, this same receptor can be found on endocrine (hormone-releasing) cells. Rather than triggering a conscious perception of sweetness, here their job seems to be to identify sugars, including those released from carbohydrates, to help to coordinate the release of food-related hormones, such as insulin, which removes glucose from the blood. They also play a role in the release of chemical signals that indicate 'satiety' or 'hunger', and so which help to regulate appetite.[14]

T1R2/T1R3 sweet receptors have also been found on cells in the colon. Here, they probably respond to compounds released by useful bacteria that play a role in digestion. It's thought they work to ensure that mostly-digested food remains in the colon for just the right amount of time for maximum possible nutrient extraction. These receptors are also present in other parts of the body and brain, including the hypothalamus – the brain's master-controller of appetite.[15] And in 2017, further evidence of the hypothalamus's direct 'tasting' of nutrients in blood was revealed.

Nicholas Dale at the University of Warwick and colleagues reported that cells called tanycytes, which are found in the hypothalamus, directly sense amino acids using the exact same umami receptor that's present on your tongue.[16] Two amino acids reacted particularly strongly with tanycytes. These were arginine and lysine, which are found in high concentrations in sirloin steak, chicken, mackerel, plums. avocados and almonds. Amino acid levels in blood and the brain after a meal are known to be a very important satiety or 'full' signal, Dale comments. Eating these foods could, therefore, help you to feel less hungry more quickly.

The critical abilities of these taste-like receptors don't end there – some of them are crucial to your health. The second type of taste-like cell found in the digestive tract belongs to the 'solitary chemosensory cell' (SCC) family. This type of cell seems to be ancient; fish have versions on their skin, which they use to 'taste' the water.[17] The particular SCC found in your digestive tract is known as a 'brush' (or 'tuft') cell, because of its bristle-like micro-villi. Under the microscope, these tuft cells look an awful lot like the taste cells in the mouth.

Their main – or even sole – job seems to be to protect you. These cells express various bitter-taste receptors, including the TAS2R38 receptor (the one that responds to glucosinolates in some vegetables and also to other bitter-tasting compounds). But in the intestine, it's thought that they detect compounds released by potentially disease-causing bacteria and parasitic worms.[18] In response, the brush cells can summon immune cells, stimulate the release of microbe-killing peptides, and even trigger the release of fluids or mucus to wash out the pathogen. In fact, recent work suggests that bitter-taste receptors around the body play a star role in protecting your health. You may well even owe your very existence to these receptors.

Bitter-taste receptors have been found in solitary chemosensory cells in the airways – in the lungs and the nose. Again, they seem to be protective, as they detect signals from bacteria and probably other pathogens, summon immune cells and trigger the release of nitric oxide.[19] A potent bacteria-killer, nitric oxide also increases the beat frequency of the cilia – tiny hairs – in the lining of the airways, to help to waft out the invader.

Bitter-taste receptors have even been found in sperm.[20] We already know that sperm cells use odour receptors to follow the chemical trail of an egg. These bitter-taste receptors are thought to help sperm to sense and steer clear of noxious chemicals, and so avoid dying.

Certain types of immune system cells, including phagocytes, which engulf and destroy bacteria and dead and damaged cells, express bitter-taste receptors, too. They can use these receptors to detect the giveaway chemical signal released by invading bacteria as they 'call' to each other to gather to form a defensive biofilm. They can, then, 'listen in' on what dangerous bacteria are doing inside you.[21]

Is the Chinese proverb 良药苦口 'Good medicine tastes bitter' a coincidence? Or do some medicines work at least partly because, by binding with bitter-taste receptors on immune cells, they mimic infection and help to stimulate an immune response? This is a question that's being urgently investigated, not least because there have been efforts to remove bitter-tasting chemicals from drugs, to make them more palatable.

Over the past few decades, taste research has focused mainly on the mechanics – *how* it happens. Now it is expanding, and one fascinating area for study is how taste-receptor activation and sensitivity on and outside the tongue might generate unconscious food drives, or conscious cravings. Cravings are not usually considered a boon to one's health. They can drive us to break our New Year's resolutions or order a takeaway . . . But understanding how taste influences our food desires and behaviour could be significant in helping us to live healthier lives.

When our bodies are lacking in sodium, a hormone called aldosterone increases the number of sodium channels on taste cells, making us more sensitive to salts.[22] But there are all kinds of examples of people eating apparently odd things. This can take the form of an eating disorder, called pica.[23] People with pica can feel the urge to consume things that are definitely not food, including ice, hair, paint or even cigarette butts. However, sometimes these cravings are for unconventional foods that do contain nutrients – such as ants. Could it be that specific nutrient deficiencies are altering the expression of taste receptors either in the mouth or the gut, which then influences such cravings?

It's possible.

Could taste-receptor signals from the gut or the brain help to account for that commonplace experience of feeling full of spaghetti bolognese, say, but absolutely having space for a slice of cheesecake?

Probably.

'It's a very exciting area where we have incomplete knowledge . . .' is Margolskee's comment.

We've learned how critical tasting is to our entire body and how important the various receptors are. But there are notable differences in how these taste receptors work. This is true both in the mouth (if you hate coriander, that's probably because you have one particular version of the gene for TAS2R50, a bitter-taste receptor)[24] and, more importantly, in our body.

Of all the receptor variations, those involved in bitter-tasting tend to trigger the biggest dinner-time arguments.

'I do not like broccoli. And I haven't liked it since I was a little kid and my mother made me eat it. And I'm President of the United States, and I'm not going to eat any more broccoli!'

This was President George Bush Sr, as quoted by the *New York Times* in 1990.[25] Given his feelings on the matter, it's pretty safe to say that Bush was what's called a 'super-taster': his genetic make-up meant that he, like roughly a quarter of his fellow Americans, had two copies of the same TAS2R38 gene variant that influences sensitivity to glucosilonates.[26]

For people with the same TAS2R38 make-up as Bush, another compound, called 6-n-propylthiouracil (often shortened to 'Prop'), also reliably tastes very bitter. Prop is the compound that is typically used in bitter-taste assessments. Some people have two copies of a different TAS23R38 variant and for them, it doesn't taste bitter at all – making it hard for them to understand why certain people hate broccoli so much. The rest have one copy of each, and they tend to fall somewhere in the middle – broccoli and other brassicas may taste bitter, but not revoltingly so.[27]

Some taste researchers argue that there's a lot more to super-tasting than this.[28] They think that people who are more sensitive to Prop are also more sensitive to sucrose (which, of course, tastes sweet), citric acid (sour) and sodium chloride (salty), and that this is because they have more of those papillae on their tongues.

If you're curious about how many you have, you can check using blue food colouring, a sheet of paper (ideally waxed paper), a hole puncher and a magnifying glass. Cut out a circle of paper with a hole punched through the middle of it. Next put a drop of that colouring on the tip of your tongue, then swish some water around in your mouth. Swallow a few times, to get rid of excess water and saliva, and look at your tongue in the mirror. The papillae should stand out against a background of blue. Now hold your bit of paper against your tongue and count how many there are in the hole-punched space. If it's over thirty, you might like to call yourself a super-taster.[29]

The concept of this kind of generalist super-taster has had quite a bit of attention. However, some studies have failed to find a clear relationship between the number of papillae on a person's tongue and their taste sensitivities, and not all taste researchers agree that there's sufficient evidence to support the idea.

Where there is general agreement about super-tasting, it's restricted to TAS2R38 and bitter perceptions. And there is an abundance of work tying genetic variations in bitter-tasting to health. This can involve tongue-taste. For example, people who are more sensitive to Prop seem to drink a bit less alcohol and coffee.[30] However, the really interesting taste links to health do not involve the mouth.[31] Because we use bitter-taste receptors to listen in on bacterial chemical chatter, people who are less bitter-sensitive seem to be more vulnerable to infection.

Some of the clearest evidence for this was published in 2014. A team from the University of Pennsylvania reported that Prop non-tasters are more susceptible to chronic sinus infections that require surgery than people who are Prop-sensitive.[32] It's thought this is because these people's immune cells are insensitive to the chemical calls of invading bacteria. 'They won't understand that there is a forming infection – so it can get well-established and cause a severe sinus infection,' Margolskee explains.

Margolskee and his colleagues have found, too, that rodent Prop non-tasters are more susceptible to gum disease, caused by bacterial infection. The next step is to explore whether this is also true for people, though it seems likely. There are even hints that genetic variations in bitter-taste receptors may play a role in cancer, but it's very early days for this research.[33]

There are also medical consequences of our genetic differences in sweet-taste receptors. These differences help to explain why some people have a 'sweet tooth' while others can (I'm told) take or leave a doughnut.[34] In fact, not everyone likes sugar, especially at high concentrations, and a pair of Monell researchers have found that versions of the TAS1R2 and TAS1R3 sweet receptor genes will predict who falls into that category.[35] The researchers also noted a relationship between bitter and sweet perceptions.

They found that children who are genetically more sensitive to bitter compounds also find very sweet compounds more pleasant, and they show a greater preference for sodas rather than milk, compared with less-bitter-sensitive kids. 'Overall, people differ in their ability to perceive the basic tastes, and particular constellations of genes and experience may drive some people, but not others, toward a caries-inducing sweet diet,' they note. In fact, there's some evidence that this can indeed happen. A group of Turkish paediatric dentists recently reported that children with particular variations in the two sweet-taste receptor genes that make sugar more appealing also suffer from more tooth decay.[36]

When paired with TAS1R1, TAS1R3 is important for umami tasting. One study investigating differences in people's sensitivity to umami found a ten- to twenty-fold variation, mostly related to common variants in the genes for these two receptors.[37] People with one particular variant of TAS1R1, for example, needed only half as much glutamate to experience an umami flavour as people with a different version. To them, a dashi, or a Sunday roast, would taste twice as savoury.

Unsurprisingly, variations in our taste perceptions shape our preferences – from having a sweet tooth to hating anything featuring coriander. The fact that taste receptors throughout our body can impact our physical health highlights the influence these receptors have on our lives. But there is also evidence that they can influence your personality.

One of the strangest links between taste perceptions and personality concerns a trait that some psychologists argue should be incorporated into the standard model, to make a Big Six. This trait is named after a man whose writings include this wonderful multi-sensory metaphor: 'Men judge generally by the eye than by the hand, for everyone can see and few can feel. Everyone sees what you appear to be, few really know what you are.' Also: 'It's better to be feared than loved, if you cannot be both.' And let's not forget this: 'If an injury has to be done to a man it should be so severe that his vengeance need not be feared.'

Niccolò Machiavelli, the Renaissance Italian writer and diplomat, is infamous for his belief that when it comes to maintaining political power, any means justify the end. As a personality trait, 'Machiavellianism' was first described in 1970. It refers to people who are willing to use cunning and deceit to achieve their goals, and who have a pretty low estimation of most people's moral fibre.

The standard scale for measuring Machiavellianism explores the extent to which you agree or disagree with statements like 'It is hard to get ahead without cutting corners here or there' and 'The biggest difference between most criminals and other people is that criminals are stupid enough to get caught.' Together with psychopathy and narcissism (an inflated belief in one's own importance, which other people must recognise, at their peril!), Machiavellianism belongs to the so-called Dark Triad of personality traits.

Across two studies of almost 1,000 Americans, Christina Sagioglou and Tobias Greitemeyer at the University of Innsbruck, Austria, have found that people who like bitter-tasting foods and drinks – who are partial to dark chocolate or strong coffee, for example – also have what the researchers describe as 'heightened sadistic proclivities'.[38] These people were more likely to score highly for Machiavellianism and also 'everyday sadism', which involves gaining pleasure from causing other people pain. (One of the tests for this was a question that reminded me distinctly of a type that my eight-year-old loves to put to his brother: Would you rather grind bugs to death, put your hand in a bucket of iced water or clean dirty toilets . . .? Everyday sadists opted to kill bugs. Or if they didn't, they later said that they regretted their decision.)

As Sagioglou and Greitemeyer observe, the results suggest a close relationship between the bitter-taste system and dark personality traits. In fact, to the researchers' knowledge, this was the first study linking taste preferences to anti-social personality traits. Overall, research connecting what people like to consume to their personality is still in its infancy, they note, adding: 'This is somewhat surprising, considering that eating and drinking are such ubiquitous and universal phenomena.'

Why might a preference for bitter tastes tie in with dark traits? Perhaps people who gain a sense of reward from what to most people is the negative sensation of causing other people pain also gain reward from the 'aversive' stimulation of bitter-taste receptors. But could this preference perhaps have any direct impact on their personality? And, the researchers also wondered, might a relatively regular stream of bitter-taste perceptions influence how they feel about and behave towards other people?

In English, there's certainly a clear association between bitterness and threat. We talk about having 'bitter enemies' and crying 'bitter tears'. In fact, as Sagioglou and Greitemeyer note, while we tend not to have strong emotional reactions to everyday sights or sounds, we automatically think of food and drink as tasting 'good' or 'bad', and while we can come to enjoy some bitter tastes, bitter is generally 'bad'.

To explore how bitter-taste perceptions might influence social perceptions in a wide range of people, they first gave groups of volunteers either a drink of extremely bitter gentian-root tea or sugary water, then gave them various questionnaires and tests. Those who'd drunk the tea were left feeling significantly more hostile (and how much they'd disliked the drink didn't affect this finding). In a separate experiment, participants who'd been given either the tea or water interacted with an experimenter on what they were told was a creativity task. The real point of this part of the study lay in the step after this, when they were asked to rate the experimenter on factors such as friendliness and job competency. Compared with those who'd had water, people who'd drunk the tea were harsher.[39]

We know that in making judgements about which house to buy, say, we unconsciously tap into our bodily states, such as our heart rate (and I'll look more at this in Chapter 12). It's been suggested that something similar happens during our judgements about other people: if we're feeling 'off', that could be because there's something 'off' about that other person. Since bitter tastes (like a strong fishy smell) trigger disgust, feelings of disgust related to bitterness in food or drink might be misattributed, leading us to feel disgusted about someone else, and their actions.

Other studies do support the idea. In one now well-known experiment from 2011, a team based in New York presented volunteers with a set of six morally dubious scenarios. These included someone eating their own dog (when it was already dead), a student stealing library books, and two cousins having consensual sex. The participants had to rate how objectionable they found each scenario on a scale from one to ten. Those who'd first been given a bitter drink rather than a sweet drink or water were much more judgemental. What's more, the researchers had asked each person taking part to indicate whether they were politically conservative, liberal, or neither, and they found that it was the conservatives who accounted for the result. While liberals were no harsher whether they'd been exposed to a bitter taste or a sweet one, the conservatives were.[40]

The findings open up a host of practical questions, such as: Should jurors avoid overly bitter drinks or foods? Could political attitudes and orientations be moderated by particular diets? The researchers also cite the Victorian-era art critic John Ruskin, who wrote:

Taste is not only a part and index of morality, it is the only morality. The first, and last, and closest trial question to any living creature is '*What do you like?*' Tell me what you like, I'll tell you what you are.[41]

In everyday language, we explicitly link physical and moral disgust. Underhand behaviour 'leaves a bad taste in the mouth'; immoral acts are simply 'disgusting'. And work published in the journal *Science* has shown that whether we taste something bitter or witness the unfair treatment of another person, our faces show the same 'disgust' response. The researchers suspect that moral revulsion at the thought of incest, for example, evolved from primal, gustatory bitter-taste disgust.[42]

There seems to be a relationship, too, between 'sweet' tastes and being 'sweet on' someone. 'Sweetie', 'honey', 'sugar' – these are all terms of endearment. And various studies have linked sweet tastes and psychological feelings of attraction or even love. One found, for example, that students who consumed a sweet drink versus a neutral-tasting drink gave more favourable ratings to fictional potential dating partners.[43] In 2019, a team in China also published work revealing that sweet tastes promote our brains' processing of romance-related words.[44] 'These findings support the embodied effect of sweet-love,' the researchers write, adding that this is a clear 'cross-modal' effect, with perceptions in a sense (taste) influencing the way we process words. It's also an effect with all kinds of consequences. It explains why chocolates and sweet-smelling flowers are perfectly acceptable Valentine's gifts, while an exotic vegetable or a tin of fermented fish might be flung back in your face.

Why do we associate sweet tastes with love, or attraction? Because, surely, sweet things and desirable partners both stimulate our brain's reward system. However, 'cross-modal' effects don't only involve links between a sensory perception and another kind of brain activity. Perceptions in one sense can also influence perceptions in the other. And it turns out that this kind of sensory cross-talk is enormously important for appreciating our experiences with food. In fact, when it comes to how we perceive what we eat and drink, examples of cross-modal perceptions are rife. And chefs with a rich understanding of how they work can turn themselves into master-manipulators.

Charles Spence, at the University of Oxford, is one of the best-known psychologists in the field of 'multi-sensory' perception – how our brains integrate information from different senses. He's particularly fascinated by how this influences our perceptions of food and drink, and his findings in this field represent a kind of smorgasbord of delightfully surprising offerings.[45] Such as:

- The same glass of Rioja is perceived as 'fresher' when sampled in a room lit with green light, with staccato 'sour' music piped in, and 'fruitier' in a red-lit room with 'sweet' legato melodies playing.
- If people eating Pringles are allowed to listen only to the higher-pitched portion of their munching, the crisps seem 15 per cent fresher. (This was the first study to demonstrate that manipulating sound alone can modify perceptions of food. It won Spence and his collaborator the 2008 Ig Nobel Prize for nutrition.)
- Coffee tastes nearly twice as bitterly intense but only two-thirds as sweet when drunk from a white mug, compared with a clear glass one.
- Being in an aeroplane dulls our perceptions of sweetness but enhances our perceptions of the intensity of umami flavours.

Traditionally, psychologists and neuroscientists have studied the senses in isolation. But work over the past few decades has made it clear that 'cross-modal' sensory effects can be very powerful indeed.

As well as running experiments in the lab, Spence works with chefs, airlines and food companies, to take his research out into restaurants, aeroplanes and supermarkets. Jozef Youssef, a young chef based in north London, is one collaborator.

Every month, Youssef serves up multi-sensory dinners to a group of ten customers. Well, to those that can find him . . . It isn't easy. Getting there from central London using public transport involves taking a tube to High Barnet, at the end of the Northern Line, then a seventeen-minute walk to a former industrial building that's now divided into units. Crumbling brick steps lead to a concertina metal gate and a sign reassuring visitors that the 'LIFT DOES WORK'.

It's a very slow goods lift with an unsettling judder. At this point, Youssef notes, most of his guests, used to slick high-end restaurants, are wondering what exactly they've signed up for. But that's part of the experience – because then there's the sensory flip of walking into a high-ceilinged white room with a sculptural screen that resembles a coral reef to one side and the open kitchen to the other.

A talk given by Spence, and then an internship at the Fat Duck restaurant in Bray, where at the time Spence was working with the chef Heston Blumenthal, opened Youssef's eyes, he says, 'to the fact that there's not only a science to the way we cook, but a science to the way we dine. Flavour,' he notes, 'is a construct of the mind. And by affecting any of the sensory modalities, we can change a person's experience of a meal.'

What Youssef can't change, of course, are genes or prior experience. He's well aware of genetic variations in bitter-tasting. All the diners are given a Prop taste-testing strip, to get them thinking and talking – and, as we talk, he points to the clear, rounded cup of coffee beside me. 'We all live in separate flavour worlds. The flavour of that coffee is built up from every other cup of coffee you've had over a lifetime.'

But he can play with some consistent and often shared ways in which perceptions in one sense affect those in another. 'Everything that Professor Spence talks about, and his research, form the underlying principles from which we design the experience,' he says.

At the time of my visit, the first course on the menu is 'coloured blobs' (Youssef's description). The four balls are green, blackish-brown, white and red. Before they taste them, diners are asked, based on appearances alone, to identify which is salty, bitter, sour and sweet.

Most guests decide that the white one is salty (presumably because salt is white), blackish-brown is bitter (again presumably because of an association with coffee, tea and dark chocolate), green is sour (perhaps because unripe fruits taste sour) and red is sweet (presumably because ripe fruits are sweet). This course was inspired by earlier research on taste-colour associations, but results from Youssef's in-house experiment fed into a joint research paper with Spence, and others, published in 2016.[46]

We learn, it seems, to associate the colour red with sweetness and fruitiness; green with sourness and freshness; white with saltiness and black with bitterness. These unconscious associations are thought to explain why the same glass of Rioja seems fruitier when drunk under red light, but fresher in green light – and why a white mug, which would accentuate the dark brown colour of coffee, can enhance bitter-taste perceptions.

Another dish on Youssef's menu is called 'Bouba Kiki', after a classic cross-modal effect.

One of the two elements of this dish is made using these ingredients: raw sea bass, lime, rhubarb, granny smith apples, vanilla and corn.

The other includes sweet potato, curd, pomegranate molasses, parmesan oil, paprika and sage.

If you had to label one set 'bouba' and one set 'kiki', which would be which?

If you went for 'kiki' for the first and 'bouba' for the second, you'd be in agreement with the vast majority of Youssef's customers. Again, research led by Spence was the inspiration. Spence has found that the word 'bouba' evokes the concept of more rounded, sweeter and fatty tastes, while 'kiki' evokes sharper, crisper, fresher flavours.

This foundations for this work date back to 1929, and a study of Spanish-speakers by the psychologist Wolfgang Kohler.[47] Kohler found that a round, blobby shape was more likely to be matched to the word 'baluba', and a spiky star-shape to 'takete'. 'Bouba' and 'kiki', which are now widely used instead, come from much later US research, which found that more than 95 per cent of both American undergraduates and Tamil-speakers in India agreed on the word-shape pairings. Further studies found that people also tend to pair carbonation, in sparkling water, with an angular shape and flat water with a rounded shape. There seems to be something consistent about the sensory associations relating to 'round' and 'spiky', and even the emotional associations – which of bouba and kiki is angry, and which is calm? I bet we'd agree.

But *why*?

It was only in 2019 that a series of studies led by Beau Sievers at Harvard University provided a unifying answer.[48] It relies on a concept called the 'spectral centroid'.

Images and sounds can be decomposed into a spectrum that contains many components of different frequencies, the team reasoned. And those components will differ, depending on the shape or sound. A shape with a few smooth curves, for example, is made up of lower frequencies than a shape with lots of straight lines and corners. The spectral centroid is essentially the average of this frequency spectrum.

Ask people (as the team did) to draw an 'angry' shape, and they'll tend to draw something very like kiki, while they'll draw a bouba-type outline for 'sad'. Sievers and his colleagues also found that when people *are* angry, their speech and movements consistently have a higher spectral centroid than when they are feeling sad or calm. A high spectral centroid characterises not only the signs of anger, then, but also spiky shapes – such as kiki and death metal logos, and brutalist building design – while a low spectral centroid is common to a bouba shape, the sound of the word 'bouba', clouds, lullabies and the swirling gravel of Zen gardens, all of which we link to feeling calm, or possibly sad, depending on the context. Sweet, 'rounded' tastes fall in easily, too, with clouds, while 'sharp' sour 'notes' sit well with their own sensory fellows.

However, although most of Youssef's diners agree on which list of ingredients should be kiki and which bouba, the colour-taste pairings show less consistency and some clear geographical variations. People from Asia tend to pair not white but black with salty, for example, presumably because of the saltiness of a local staple ingredient, soy sauce. But like Pavlov's dogs, which came to salivate at the sound of a bell that had, previously, repeatedly been rung in conjunction with food, years of eating sweet, red strawberries and other fruits can lead us to perceive sweetness when we eat a red-coloured food. In the same way, perhaps, we perceive textural 'crispness' when we hear the high-pitched sounds that we've learned to associate with munching into something fresh, such as an apple.

As with those Yale study participants who came to hear a tone whenever they were presented with a chequerboard image, consistent multi-sensory experiences build very strong flavour expectations. So strong, in fact, that they can lead us to perceive things that aren't there – and not to perceive things that are.

At the ancient University of Bordeaux, which lies in the heart of one of the world's most famous wine-making regions, oenology students learn everything from how to grow vines to the sensory qualities of various grapes. In 2001, Frédéric Brochet at the university's Faculté d'œnologie gave fifty-four of these undergraduate students a test.[49]

They were each presented with two glasses of vintage Bordeaux. One was a red (a cabernet-sauvignon and merlot blend). The other was a white (semillon and sauvignon grapes). The students were allowed to sniff and taste the wines, then they were asked to match items from a list of descriptors to each. As might have been expected, they tended to choose words such as 'honey', 'lemon' and 'grapefruit' for the white wine and 'prune', 'blackcurrant' and 'chocolate' for the red.

A week later, the same students were again given two glasses of wine. However this time, while the white wine was the same as before, the 'red' was a glass of the white, coloured red with an odourless dye. The students chose the same 'white wine' terms for the white as before, but 'red wine' terms, like blackcurrant and chocolate, to describe the 'red'.

They were just students. Perhaps, knowing what the descriptors 'should' be, they consciously overrode whatever their senses of smell and taste were telling them in favour of classic red/white odour terms. It's possible. But Brochet thinks that their conditioned expectations of what a red-coloured wine should be like were so strong that signals from the eyes over-powered those that we trad-itionally think of as being responsible for perceiving flavour – and they smelled and tasted what they saw.

In this case, the students' perceptual 'best guess' of what they were tasting wasn't up to much, as prior knowledge had swamped reality. One of the biggest perils to our ability to smell/taste wine accurately, then, is ourselves. But when it comes to the taste system itself, there are other, bigger threats.

You'll be familiar, I'm sure, with the persuasive argument that so many of us in the West are overweight because the drive to consume sweet high-calorie sugars – a drive that so aided our ancestors in their fight for survival – no longer serves us well. In a world where food is easily available, it's correspondingly easy to eat too much of it. And it's not exactly hard to find meals and snacks packed with sugars as well as salts and fat. Our modern diets, combined with lack of exercise, are making us fat – and that itself is bad news for taste.

It's been known for some time that obesity blunts taste. In 2018, a US group came up with the most convincing evidence to date as to why: low-level, body-wide inflammation, caused by being over-weight or obese, upsets the normal balance of taste cell death and renewal.[50] The net upshot is fewer taste cells. Because we gain a feeling of reward from eating – and taste signals to the brain are critical for that – it's thought that someone with a weaker sense of taste has to eat more to gain the same level of reward that a person with a healthy taste system would enjoy. Obesity and poor taste are clinched, then, in a downward spiral, encouraging and worsening each other.

However, there is also evidence that people who are significantly overweight or obese and who lose weight can enjoy a taste restor-ation, presumably via a normalisation of taste-cell renewal. But even if you're a healthy weight, it's sensible to be careful about how much sugar you eat. There's some evidence that a high-sugar diet reduces the response of sweet-taste receptors. This would make precisely the same slice of cake taste less sweet, which could encourage you to eat even more of it.

Conversely, a low-sugar diet can make a sweet food taste sweeter, a change in sensitivity that helps you to identify foods that contain sugar, but also potentially means that you'll need less to make you feel that you've had 'enough'. I have first-hand experience of this. Diagnosed with gestational diabetes while pregnant with my second son, I had to cut out all processed sugar and limit my consumption of fruits. After a while, I noticed changes in how foods tasted. A strawberry became *insanely* sweet.

For me, drastically reducing my sugar intake did make a big difference to how sweet foods tasted. But taste-chemical deprivation aside, it may not be possible to train your sense of taste – at least, not in the way that you can train yourself to be more sensitive to smells, say. 'Smell is a more flexible system than the gustatory system. Taste is pretty hard-wired,' observes Beverly Cowart at Monell.

Doing your best by your sense of taste means, then, protecting it, by being careful about what you eat. But it also means, surely, celebrating all the very many ways that the taste receptors throughout your body look after you – guiding you to the good things, while also keeping you safe from harm.

5

Touch

How to climb a mountain with your tongue

Touch is not a single sense but many.

Aristotle, *De Anima*

Picture a Venus flytrap lying in wait in boggy ground in a longleaf pine forest in South Carolina. As an unwitting grasshopper lands on one of the flytrap's ruby lobes, it brushes against a sensory hair – but all is not lost. If the grasshopper can leap away without disturbing a hair again, it will be safe. But just one more brush – and the trap springs shut. Now, the plant keeps on counting.

Touch one meant likely contact. Touch two: contact with potential prey confirmed – spring trap! Touch three and four: start to make digestive enzymes. Touch five: intensify digestion; nutrients from the liquifying meal can now be sucked up.

Using touch signals alone, the flytrap can time the stages of its deadly response. It can also estimate the size of its prisoner, and so ensure that it releases enough in the way of digestive enzymes to make a good meal of it, but not so much that those precious compounds go to waste.[1]

For the Venus flytrap, its sense of touch infamously allows it to prey on insects and spiders. But at its most basic, touch provides a life form with a sense of where its body ends and the rest of the world begins. This is fundamentally important information. Unsurprisingly then, everything from bacteria to seahorses and mosses to tapeworms register touch. In fact, Aristotle opined, it is the most necessary of all senses, since an animal without it would not be able to stay alive.

In *De Anima*, Aristotle argued that touch is not a single sense.

About that, he was certainly right. However, that was because for him, perceptions of heat and cold, for example, were part of our touch experience. Now we know that these are separate senses. But we also know that touch is itself complex.

For humans, touch has evolved to have two individual aspects. First, there's discriminative touch. This is essentially a more sophisticated version of what the Venus flytrap, or pretty much any plant in your garden, enjoys. It is practical touch, informing us about what's in contact with us, where, and whether we might need to adjust our muscles if we don't want to drop that glass, or slip on that rock. But even this branch of touch is not really one, but two senses. Using separate sensory hardware, we can feel pressure and also vibration.

The second branch of touch is different. When you gently stroke your baby's cheek, or your lover slowly brushes your neck, a distinct population of touch sensors in the skin react. This is so-called 'emotional touch' – and it's the kind we get from other people. Properly described only in the 1990s, it is essential for our healthy development and our wellbeing. In a world where millions of people live alone (7.7 million just in the UK), it's also a type characterised by desperate levels of deprivation.

Understanding touch in all its forms, then, is vital to both our physical and emotional wellbeing. Before we get ahead of ourselves, though, we need to start at the beginning – what happens when we come into contact with something else?

Think for a minute about the kids' game in which you're asked to close your eyes and identify an object using touch alone: grape, piece of Lego, playing card, teaspoon . . . Easy.

It was a German scientist, Johannes Müller, who, in 1842, suggested that qualitative variations in perceptions might be underpinned by the activation of distinct receptor types.[2] And when it comes to what you can touch with your fingertips, it turns out that four kinds of mechanical force sensor allow us to tell apart a sandcastle from silk, or a teaspoon from a grape.[3]

Physiologists divide our skin into two main types. There's the 'hairy' type that covers much of our bodies, whether it looks especially hairy or not. Then there's the hairless skin of our finger-tips, palms, toe-tips, soles of our feet, lips, nipples and clitoris, and the foreskin and head of the penis. This skin goes by the un-appealing name of 'glabrous'.

The two types of skin have different touch roles. Glabrous skin in the fingertips, especially, is specialised for working out what something is and what it isn't. In the mouth, it allows us to detect the location and texture of whatever it is that we're eating, as well as the position of our tongue, which is essential for talking. In the penis and clitoris, it is fine-tuned for responding to sexual types of stimulation. A primary role of mechanoreceptors in hairy skin, meanwhile, is to inform us when something else is in contact with us: roughly where on our body the contact is happening and what it's likely to be – whether it's a shower of rain, an ocean breeze, a hand, or the flick of a snake's tongue.

Especially densely packed in the skin of the fingertips and also nipples, clitoris and the glabrous skin-parts of the penis are 'Pacinian corpuscles', named after the nineteenth-century Italian anatomist Fillipo Pacini. Under a microscope, these receptors are multi-layered, almost onion-like. They consist of a single nerve fibre wrapped in layer after layer of supporting cells, and they are highly sensitive to even tiny vibrations.[4] For an example of Pacinian corpuscles in action . . . let's go with the fingertips. And you, reaching for a wine glass from your kitchen cupboard.

The faintest, tiniest drag of your finger against the glass would stimulate Pacinians in the skin in your fingers, informing you that you are indeed holding a smooth, sheer object, rather than a wooden cup, say. As the neurobiologist Gary Lewin, at the Max Delbrück Centre for Molecular Medicine, in Germany, explains: 'If you move your finger slowly across a surface, the roughness of that surface as well as the speed of your finger means that you are vibrating Pacinian receptors as your finger moves.'

No, we didn't evolve to be able to identify wine glasses with our eyes shut, but our ability to hold objects and use tools is crucial to our survival. And these receptors, which are sensitive both to deep pressure and to high-frequency vibrations, can make you feel that a solid tool is essentially an extension of your own hand.[5]

Now imagine that your fingers slip imperceptibly slightly on the wine glass (or tool). Slip-sensitive Meissner's corpuscles, close to the surface of your skin, react instantly. When these bulbous structures are squashed, even very slightly, sensory signals racing off to your spinal cord trigger a reflex adjustment to the degree of contraction of the muscles in your fingers. This causes you to subtly tighten your grip, so the glass doesn't drop to the floor.

A third receptor comes into play, too. In 1894, an Italian anatomist called Angelo Ruffini first described his eponymous endings, which he initially called corpuscles, in cat skin. (Ruffini even used himself as an experimental subject; a contemporary reported having a microscope slide of these corpuscles in a segment that Ruffini had dedicatedly, if gruesomely, cut with a scalpel from his own arm . . .) Ruffini endings report on stretch, detecting, for example, stretch in the skin of your palm, caused by the pressure of the glass on your hand. They connect to a kind of 'slowly adapting' nerve fibre, which keeps on sending signals for as long as the stimulus is there.[6] (Pacinians, in contrast, are 'rapidly-adapting', sending messages only when their stimulation changes.)[7]

Then there are Merkel cells (sometimes called Merkel's discs),[8] which can be found in the skin of all vertebrates, and which were first described in 1875 by a German, Friedrich Merkel. (He called them *Tastzellen* – touch cells).[9] In your fingertips, they're found in clusters of as many as 150 at a time. They are incredibly responsive to pressure, allowing you to detect the corners and edges of an object, and also its texture. The merest brush of your finger across the rim of the wine glass would stimulate these cells. In fact, if you were to close your eyes and be given same-sized and -shaped sections of, glass, wood, metal and plastic, you'd probably have little trouble working out which is which; in part, this would be because these materials draw heat away from the finger at different rates – but signals originating in Merkel cells would be critical. This class of cell, which connects with slowly-adapting sensory neurons, underpins the fine discriminatory ability of your fingertips.[10] (The fact that they are far less densely packed in some other regions of glabrous skin explains why, as David Linden at Johns Hopkins University in Baltimore notes, it's possible to read Braille with your finger pads, but not with your genitals . . .[11])

Only recently, it was discovered that our fingertips are even more exquisitely sensitive than scientists had realised. In 2017, researchers at the University of California, San Diego investigated whether, by touch, volunteers could tell the difference between silicon wafers that looked identical but which had a top layer either of mostly oxygen atoms or of fluorine and carbon atoms. When given two wafers of one type and one of the other and asked to run their fingers lightly across them, the volunteers correctly picked the odd one out 70 per cent of the time. We can, it seems, detect differences in surfaces that are just a single molecule deep.[12]

Thanks to particular groups of neurons in the somatosensory cortex of your brain, where touch signals are processed, you can also instantly respond to different aspects of texture. In 2019, work at the University of Chicago revealed that while some of these groups respond to roughness, for example, others respond to finer features or specific patterns of indentation of the skin.[13] 'Velvet is going to excite one subpopulation of neurons more than another, and sandpaper is going to excite another overlapping population. It's this variety in the response that allows for the richness of the sensation,' explains Sliman Bensmaia, who led the work.

Incidentally, since the way the brain interprets skin touch is built up from long experience, this can make for some interesting touch illusions. One is even named after Aristotle, and it's easy to try for yourself. All you need is a small, spherical object, such as a frozen pea. Cross your fingers, position the pea so that you can touch it with both fingers, close your eyes and touch it. It will probably feel as though you're touching two peas. Why? Because the outsides of two fingers are touching one object at the same time, while experience (with uncrossed fingers) is maintaining that for this pattern of signalling to happen, two separate objects must be there.[14]

But if our finger-sensing is easily confused, it's even worse for our toes. In fact, even if you keep your toes in their regular positions, it's surprisingly easy to be wrong about which is being touched, according to research led by Nela Cicmil at the University of Oxford.[15] Cicmil's team first got people to close their eyes and hold their hands out, then touched them on one finger. The volunteers could correctly identify which had been touched about

99 per cent of the time. Now, if you can find a friend with a pencil to help, strip off any shoes and socks, and ask them to touch your toes, one at a time, in any order. When Cicmil and her team tried this with their volunteers, if it was the big toe or little toe that was being touched, they got that right about 94 per cent of the time (which means of course that they were occasionally wrong about even this). But for the middle toes, moving from the big toe out, they were right only 57, 60 and 79 per cent of the time, respectively. No one got it right 100 per cent of the time. And for most, the two toes beside the big toe were especially confusing. Why?

'We have suggested a model in which rather than sensing each toe separately, the brain just sees five blocks,' says Cicmil. However, 'the gaps between the actual toes do not correspond to the boundaries of those blocks.'[16] This, then, seems to be the cause of the confusion. Well, that and our habitual wearing of shoes, which deaden sensory input to our feet.

In hairy skin, with the exception of Meissner's corpuscles, we have the types of mechanoreceptor that are found in glabrous skin, but not at the same densities (which is why we are far better at working out what something is using our fingertips rather than the back of a forearm). But in hairy skin, we also have the touch asset of specialised sensory nerve endings that wrap around the base of the hair follicles. When a hair is bent, signals are fired off to the brain at a rate determined by the degree of bending. Using the rate and pattern of these incoming signals, your brain can differentiate between the brush of wind and the crawling of a spider. (If you depilate, you'll be familiar with the relative numbness that results from the loss of these hairs.)

Differences in the densities of particular touch receptors can mean that in some parts of our body, even with hair, we are relatively numb.[17] The skin on your chest and back contains a hundred times fewer touch receptors per square centimetre than your fingertips. If you can find another pencil for your friend, you can explore this for yourself. Close your eyes, and ask them to hold the pencils together and gently prod you at a point on your fingertip. Then ask them to move the pencils further apart gradually, and try to notice when you can first distinguish two distinct touches. On average, a person can sense two separate pressure points on a fingertip when they are between two and four millimetres apart. On the back, this dead zone can stretch to three to four centimetres.

Still, your relatively Merkel-impoverished back, neck and forehead are rich in a distinct type of touch fibre that we haven't met yet. These are slow-transmitting 'C-tactile fibres', and it's these fibres that respond specifically to slow-moving, gentle touch by something that's at about thirty-two degrees centigrade . . . typical skin temperature.[18] Since this type of touch generally comes from another person, these fibres, which were identified only in the 1990s, have been nicknamed caress sensors.[19] They're used not for discriminative touch, but for 'emotional' touch.

Your caress sensors send information not only to the somatosensory cortex (which handles the practical aspects of touch) but also the insular cortex, which is involved in emotion. It's a type of touch that certainly seems to be related to the grooming contact observed in rhesus monkeys, for example. Because (if it's wanted) it feels good, it encourages us to be close to other people, spend time together and to bond. In fact, it's nothing short of a 'biological necessity', argues Francis McGlone at Liverpool John Moores University.[20]

Anyone who's ever been around a baby knows that it is human touch – that infant up close, on bare skin – that is most calming.[21] Nothing says 'I am there for you' like the squeeze of a hand or a hug. And, argues McGlone, this kind of intimate, encompassing touch, which starts in the womb, is critical for the development of a healthy social brain. It also helps premature babies to thrive – to gain weight more rapidly and to be discharged earlier.[22] This discovery was made back in the 1970s. Yet it would be another twenty years before the emotional touch system in the body and brain was discovered.

It seems remarkable that so fundamental a system should have gone unrecognised for so long. This reflects historical interests in sensory biology, says Gary Lewin. In the 1960s, '70s and '80s, the vast majority of sensory research was on vision. 'We focused on the visual system because it's very clear that humans are very visually-guided,' he says.

'But I think that focus led us to under-estimate the importance of touch.'

Touch is so important that differences in how sensitive we are to it can have profound implications for everyday life. Someone who is less sensitive to touch is unlikely to be able to wield a scalpel well, for example. As a child, they might also find it harder to learn to write with a pencil at school. And, since precise touch inside the mouth is critical for our ability to speak, variations in touch sensitivity might explain why some young children take longer to learn to talk than others.

Among people with healthy touch, Lewin has found an extraordinary range of sensitivity.

In testing any group of 100 people, using vibrations at a frequency of about 125 Hz, which stimulate Messiner's corpuscles, his team always find a few who can feel vibrations measuring just 300 nanometres at their peak. However, at the other end of the spectrum, there will also be a few who struggle to feel a vibration of three *micro*metres – a vibration that is ten times bigger. Everyone else falls somewhere in between, forming a bell curve. 'We don't really know why that is,' Lewin admits. 'It could be to do with the receptor endings – some people just have more sensitive sensory neurons. Also, it's possible that their nervous system is better at detecting the signals that come in from the skin.'

Lewin and his colleagues have studied identical and non-identical twins to explore the role of our genes in influencing touch sensitivity and also hearing acuity (both of which rely upon mechanoreceptors). They found that both are highly heritable. About 40 per cent of your touch performance is attributable to your genes. The team also noted that those people in the study with excellent hearing tended to be highly sensitive to touch, too. The reverse was also true: those with relatively poor hearing weren't as touch-sensitive.[23]

How can we understand what causes those differences?

Like a person sitting in the stand of a football stadium, a mechanoreceptor in the skin is surrounded by other cells, but also supporting structures. In the stadium, there are levels and seats. In the skin, the cells are, to use Lewin's phrase, 'kind of glued together' with what's called 'extra-cellular matrix'. Lewin and his team have shown that this matrix is critical for the opening of force-sensitive (mechanoreceptive) ion channels. After all, something must connect the matrix to the channel – and Lewin has evidence that a protein called Usher2A is involved: 'Picture a sink full of water, with a plug and a chain on that plug. You can think of that plug as the ion channel. When you open up that channel, ions flow into the cell, and excite it.' Lewin thinks Usher2A is a protein in the matrix that can pull on that chain. Certainly, he's found that people with particular mutations in the gene for Usher2A are not only less touch-sensitive but also congenitally deaf. (And as we know, force-sensitive ion channels in the connecting fibres between hair cells in the cochlea are critical to our ability to hear.)

Pioneering research based out of the lab of Ardem Patapoutian at Scripps Research, La Jolla, California, also involving Lewin and his team, has revealed that another protein called Piezo2 is very important to our sense of touch.[24] Piezo2, it turns out, is the critical ion channel in Merkel cells. Physical force alters its shape, allowing sodium or other positively-charged ions to flow in. This triggers an electrical impulse along the sensory neuron to the spinal cord and ultimately to the brain. People who lack functioning Piezo2 can feel slow, brushing touch on hairy skin, but they struggle to register light touch.

In 2016, the first two people with precisely this Piezo2 mutation – a girl aged nine at the time and a woman aged nineteen – were described in a journal paper by a US team.[25] One of the girls found it impossible to tell, without looking, whether one or both small tips of a pair of callipers were being firmly pressed into her palm. Although they both felt gentle brushing of hairy skin, one said that it was a prickly sensation, rather than pleasant.

In theory, it may become possible not only to help people with faulty Piezo2 to feel touch, but for people with normal touch to get even better at it. Lewin's team recently identified a drug that can shut touch reception down. As a point of biological principle, if you can specifically inhibit touch, then the opposite – using drugs to enhance touch – must also be feasible, he says. Also, other teams have found molecules that activate an ion channel that is related to Piezo2. 'If you can do it for a related channel, you can probably do it for Piezo2,' Lewin comments. 'No one has yet designed drugs that can sensitise your touch, but the bottom line is that it is possible.'

Though such drugs aren't here yet, it is possible to train touch. You can even train one fingertip (or rather, your brain) to have sharper touch responses – and, in what seems initially like a slightly spooky inter-connection, the benefits will spread to some other fingers.

Vanessa Harrar at the University of Oxford and her team first tested the ability of a group of participants to discern slight differences in two distinct touches to each of their fingertips. Then they repeated the touches, over and over again, on just one fingertip. Each time, they asked the participants what they could feel, and gave feedback on whether they were right or wrong. Gradually, these volunteers got better – their touch sensitivity for this particular fingertip improved. But when each of their fingertips was tested again, the team found that the adjacent fingers and the same finger on the other, untrained, hand had also got better. In fact, they showed similar improvements to the trained finger.[26]

How did this happen? Because, the team thinks, touch signals from those fingers go to the same region of the somatosensory cortex. When this region gets better at processing touch signals, the benefits are shared.

Perhaps even more remarkably, another team has found that repeated touch to the tip of the right index finger improves touch perceptions on the lips.[27] This is thought to happen because the region of the primary somatosensory cortex that processes touch on the face is adjacent to the region that deals with the hand – but this is a slightly blurry division, so the benefits of improving responses in one region leak a little to the other. (In fact, shifts in neural connections in these adjacent hand/face regions after the loss of a hand, through injury or amputation, can make some patients feel that when they're touched on the face, that touch is actually happening to their missing 'phantom' hand.) If repeated touch to the right index fingertip improves touch sensitivity on the lips, I can't help but wonder whether people who spend half their waking lives tapping on a keyboard make for better kissers. Perhaps you could conduct your own personal study . . .

Touch can, then, clearly be trained. And of course, at a basic level, that training starts even before birth. It takes time, though, for a child to hone these senses. The more practice they get at manipulating different objects, the better they get at touch. But we are living in a world where screens are (to a varying extent) replacing traditional physical toys, and this worries some researchers. Some also worry that touch between people is not as acceptable in many cultures as it used to be. There are, of course, some extremely good reasons for this. But one upshot of a discouragement of touching between teachers and pupils, or colleagues in the workplace, not to mention the rise of relationships conducted online rather than 'in the flesh', is a reduction in this human contact – in human touch. Since it's so important for feeling socially connected and supported, some in the field are concerned.[28]

Just how important it is – and how hard it is to live without it – was a surprise discovery for a British composer who, in the course of her work, found herself talking to people who hadn't been touched by another human in years. Their stories touched her deeply. There's just no other way to put it. And as she describes them to me, it's easy to understand why.

Steph Singer is the young, energetic creative director at BitterSuite, a group that creates multi-sensory symphonies. These symphonies are designed to engage not just hearing but also taste, smell and touch. Someone attending a BitterSuite production will be exposed to various scents and tastes, and they will be paired with a guide – a performer who will take them through physical movements while the orchestra plays.

In a workshop space in east London, I get a feel for what this is like. Singer gets me to sit on a high stool and asks me to close my eyes. Then she places one palm flat on my upper chest and the other palm on my back. It's a deep, containing kind of contact. As I stand, she shifts to hold me around the waist, so we are beside each other, pelvis to pelvis, her hands around my hips, now guiding me as we walk. With my eyes closed, I'm focused on her touch, and her (gentle, consensual) control over my movements, my body, even my will, subjugated to hers.

Singer soon noticed that during performances, these physical partnerships between concert-goers and their guides could become quite intense. At first, the group didn't have a particular focus on touch in the sessions. 'But then we realised it's an incredibly intimate communication device,' she says. 'It means you can build these non-verbal relationships between two strangers in a room, then at the end you have this eruption of shared experience. People want to talk about what they just felt, and they want to go to another audience member and find out what they experienced. Everybody gets the identical choreography, but every body experiences it differently.'

Singer was particularly struck by the emotion of people who said it was the first time they had been touched by another person for some time. At one concert, a thirty-eight-year-old man revealed that it was the first time he had been touched for seven years. 'We had another woman come, who was in her early 30s,' Singer recalls. 'She said she'd just moved to the city and over the past eight months, she hadn't really been touched by anybody. She talked about how everyone hates packed tubes, but she started to lean in ever so slightly towards people on the tube, because she just missed that contact . . . I'm not saying that everybody should be touched all the time. But contact is *human*.'

In collaboration with a researcher at King's College London, Singer went on to explore her concert-goers' attitudes to touch in more detail. At the end of a performance, they were given questionnaires to fill in. Some of the responses to 'Have you ever missed touch in your life? If so, how did that make you feel?' were very frank and unforgettable.[29] For example:

> Yes. It made me feel less fluid in my body, more rigid. I felt an ache for touch and I remembered (or tried to) what touch was like. I remembered how my mum would stroke my hair and ear when I was sick or upset.

> Yes, very lonely but I didn't realise until I was touched (at the end of a yoga class). I cried.

There are cultural differences, of course, in the extent to which it's typical or deemed acceptable to touch each other. And some researchers think these differences help to explain variations in other types of social behaviour.

When Tiffany Field, head of the Touch Research Institute at the University of Miami, observed nursery-school children in the play-ground and adolescents in McDonald's in Miami and also in Paris, France, she noticed some clear discrepancies. When it came to the nursery-school children, the French children were touched more by their parents than the young children in Miami. The French teen-agers, she noted, touched, hugged and stroked each other more than the American teens. The French teens were also less aggressive with each other, both verbally and physically. Of course, other factors could help to explain this, but Field believes that differences in the way the two nationalities touch each other is important.[30]

Based on her years of observing people touching each other in public places, she also thinks children are much more touch-deprived than they were before smartphones came on the scene. It might be tempting to give a young child a phone as a pacifier, but then they're not interacting physically with their parent, or a sibling. In fact, the ubiquity of technology means we have to be conscious of ensuring that children get plenty of human touch, she argues: 'I think parents have to make a special effort to provide as much touch as they can.'[31]

At the moment, touch isn't a sense that works at a distance. To feel all the little bumps and indents of an orange, say, you need to have it there, in your hand. But that could soon change. The video game industry is driving much of this research. What these companies would like is for a gamer not only to see and hear their virtual world, but to feel it, too. At the Reconfigurable Robotics Laboratory at the Swiss Federal Institute of Technology in Lausanne, a team of engineers is working on one approach to enabling this. They are developing a soft, artificial synthetic skin, made from flexible plastic and studded with tiny air pockets, each of which can be inflated and deflated many times a second.[32] Someone wearing this skin, as a glove, or even over their entire body, could *feel* a punch in the chest, or a gentle stroke.

Other teams are looking at developing more sophisticated artificial mechanoreceptors,[33] and even an artificial touch neuron, which can detect how hard it is being pressed and pass on that information, to control real muscles.[34]

Some groups are focusing on new ways of using touch. Just as it's possible to put hearing to an unusual use – to echolocate in the style of Daniel Kish, for example – it's proving possible to co-opt touch into other roles. And some of the results are extraordinary. For example, haptic devices that convert video images from body-worn cameras into distinct, fizzy patterns of stimulation on the tongue have allowed blind people to climb mountains. Erik Weihenmayer, an American who lost his sight as a teenager due to an inherited condition, has used a device just like this for outdoor climbs around Utah and Colorado. (Weihenmayer has even climbed Mt Everest, though for that ascent, he used 'only' touch, plus a guide.)[35]

But, some researchers argue, it's possible to hack your senses of touch, to enable new 'senses'. 'Our experience of reality does not need to be constrained by our biology' is the sci-fi-esque promise of David Eagleman, a neuroscientist at Stanford University.

Eagleman's lab initially conceived of the vibrating VEST (Versatile Extra-Sensory Transducer) as a sensory substitution device for deaf people. It's worn like a gilet, and incorporates vibrating motors (thirty-two at the last count). The idea was that when somebody spoke, their words could be translated into patterns of physical-touch stimulation. But the work has since expanded into all kinds of other areas.

In theory, it can translate practically whatever you like – stock market trends, for instance, or your blood pressure, or light outside the visible spectrum – into patterns of touch stimulation that we humans, wonderful pattern-spotters and learners that we are, can come to interpret. This takes quite a bit of conscious effort at first. But with practice, Eagleman says, these associations become automatic 'and a new sense is born'.[36]

Would this really count as a new 'sense', though? Or just a way of receiving new information via touch? People who are fluent users of Braille may lose their conscious awareness of touch signals, and come effortlessly to 'sense' the written word – but they are still using touch.

If, for argument's sake, you were to develop a new 'sense' of the stock market, say, as you already know, this would not be your 'sixth sense'. We do undeniably have other natural senses, beyond those of sight, hearing, smell, taste and touch. And it's time now to move beyond that old, limited Aristotelean model, into the extraordinary, wider world of your incredible capacity for sensing.

Part Two

The 'New' Senses

Part Two

The New Senses

There is no other sense beyond the five . . .

Aristotle, *De Anima*

As we know, Aristotle had his reasons for concluding that we have five, and only five, senses. To reiterate, he was a remarkable biologist. One renowned nineteenth-century anatomist even went so far as to declare: 'Zoological Science sprang from his labours, we may almost say, like Minerva from the Head of Jove, in a state of noble and splendid maturity.'[1] However, research since his time has expanded his little cluster of senses into a dazzling universe.

It turns out that there's a vast sensory world out there to explore – a world of little-known senses that allow you to feel everything from agony to ecstasy, and that take your body places that touch or vision could only dream of. These senses, just as much as the ones we've already met, make us who we are. They even go right to the literal heart of the human experience. But we're going to start by getting under your skin.

6

Body Mapping

How to be a prima ballerina

Stand up. Close your eyes. Raise your right hand and bring your index finger close to your forehead, but don't touch it. Now take a step forwards, and bring your hand up to connect, palm-side down, with the top of your head.

Did you manage it? If you did, of Aristotle's five senses, only touch was playing a part, telling you when your hand made contact with your skull and when your feet had lost and re-made contact with the ground. But without a mental map of the location of your various body parts, you'd have had no idea when your finger was close to your forehead – and when you tried to bring your hand back to your head, your arm would have been flailing. You'd have been lucky to get that far, however. Deprived of information about muscle activity in your legs, when you tried to walk, you'd have ended up in a heap on the floor.

We humans rely heavily on vision to help us to coordinate contact with objects and to move around. However, without vision, these things are still readily achievable. Lose your sense of where in space your body parts are located, though, and it's a very different story. In this chapter, we are going to learn how this vital sense, known as proprioception, works, and how we can use it better, to affect our physical bodies and also our minds.

For a 'new' sense, it isn't really very new. The fundamental science of how it works has been known for well over a century. From the late 1860s, the English anatomist Henry Bastian had argued for the existence of a muscle sense, which he called 'kinaesthesia' (from the Greek *kinein*, to move, and *aisthesis*, sensation) – the sense of motion of the limbs. He believed that it had to play an important role in the brain's control of bodily movements.[1] But no one knew what might actually provide that sense.

There was keen interest, though, in working it out – and the scientific momentum to do it. In 1876, the Physiological Society was founded. Leading figures proclaimed that findings from research in anatomy, experimental physiology and clinical medicine really ought to shed light on each other. And it was starting to become apparent that different regions of the cerebral cortex – the outer layer of the brain – had specialised functions. By 1881 there was even talk of a so-called motor (movement) area or 'sensori-motor area' of the brain.[2]

There were German physiologists who maintained that the brain's monitoring of movement commands despatched to muscles generated the perception of motion. But what if someone moves your arm for you, lifting it up above your head? You still feel your arm in motion and you know where it is. As the British physiologist Sir Charles Scott Sherrington pointed out, even when we're still, we have an instinctive, deep-seated sense of the position of our various body parts. Sherrington was determined to understand how.

Sherrington developed the idea of a 'muscular sense', a fundamental component of what he soon came to term 'proprioception' (a combination of the Latin for 'one's own' and perception).[3] He, and fellow sensory pioneer Angelo Ruffini (he of the Ruffini endings that detect stretch), along with many other physiologists, worked tirelessly, pushing the boundaries of the known sensory world. What they were discovering in terms of the receptors themselves, and how they despatched their messages, was remarkable. It was like pulling jewels from the flesh. (As we know, Ruffini even pulled some from his own.)

After moving on to the universities of Liverpool and later Oxford, Sherrington immersed himself in research on the connections between muscles, spinal cord and brain, and was later named joint winner of the 1932 Nobel Prize in Physiology of Medicine. Finally, through his research, Sherrington provided definitive evidence that sensory signals coming into the brain from around the body feed into posture and movement control.[4]

Perhaps you haven't thought of yourself as a 'segmented inverted pendulum' before . . . but, with your spine made up of individual vertebrae and your heavy head, that's what you are.[5] Such an inherently unstable system requires near-constant posture adjustments to keep it upright. Sherrington's research showed that reflex pathways – which enable automatic reactions that require no will – largely handle this job.

However, whether you're unconsciously staying upright on a bus or climbing Mt Everest, both tasks depend on proprioception. This, Sherrington wrote in 1906, is 'the perception of joint and body movement as well as position of the body, or body segments, in space.'[6] Essentially, then, it's the sense of when, and how fast, parts of your body are moving, and where they are in relation to each other. Are you sitting down? If you are, close your eyes. If you can *feel* instinctively that your legs are located lower than your chest, that's proprioception. If you now raise one arm, and you can feel it move, that's proprioception, too.

Admittedly, it's not exactly an accessible term, and it is a little-known sense, but that doesn't mean it's unimportant. It's like an employee who keeps everything going, but is granted only the scantest attention. Just because you rarely notice that person going about their business doesn't make them dispensable, but criminally under-appreciated. In fact, we use proprioception all the time.

Right now, for me to be able to type, my brain needs to know exactly where my fingers are in space and in relation to each other. When I first learned to type, I focused on what I could see, watching where my fingers were going on the keyboard. Because vision hogged the conscious limelight, as so often happens, I wasn't aware of the relatively subtle position signals coming from my hands and fingers. But if I close my eyes now, and carry on typing, I can feel them.

Now think back to when you learned to catch. I bet that you, like me, were told to watch the ball, not your hands. That advice works because, thanks to proprioception, your brain knows where your hands are in space. The unknown is the location of the ball, which vision can help with. To catch something, you don't need to be aware of your hand-position signals – but they are there all the same.

Three distinct groups of receptors, embedded within our bodies, along with those of all other mammals, provide us with this sense.[7] Like those essential for touch and hearing, they are all 'mechano-receptive' – they react to physical stimulation, like squashing or stretching.

Studded through your muscles are what are known as 'spindle' stretch receptors. These tiny capsules of connective tissue contain small, specialised muscle fibres that are in very close contact with the endings of sensory neurons. These neurons transmit messages that tell the brain when a muscle starts to stretch, how quickly it's stretching, and when that stretching stops.

Muscle spindles, as an entity within the body, were discovered in 1851. But it was Angelo Ruffini who, in 1892, while at the University of Bologna, first realised that they are a form of sensory receptor. Incidentally, although Sherrington was fully plugged in to the new sensory scene, Ruffini was not. His job was to do research into and teach the microscopic anatomy of bodily tissues. Through this work, sensory receptors in the skin and muscles became a passion (as did amphibian embryos, but that's another story).

When Ruffini worked on these receptors, he did so alone, and, for a long time, in relative obscurity. It was a struggle to get his breakthrough papers published. Sherrington, though, was a fan. For him, Ruffini was the go-to guy for anything in this area. In fact, when a neurologist friend of Sherrington's reported that a patch of skin on his arm had 'limited' sensation, Sherrington immediately wrote to Ruffini, asking, rather breathlessly, if he'd examine a sample:

> Kindly let me know at once if you can do this . . . And write me the exact fluid in which to place the piece of skin removed. Also say whether you would wish a single tiny snip or several, and to what depth . . . I feel sure that you are the best investigator in all the world for this enquiry.[8]

The pair's letter-writing friendship (sadly, they never met in person) had been sparked, though, not through skin receptors but through Ruffini's work on muscles. A year after Ruffini declared muscle spindles to be sensory receptors, Sherrington reported that he had traced nerve fibres from these spindles all the way through to a dorsal root of the spinal cord – the point of entry for sensory messages – confirming that they were, indeed, feeding into our sensory experience. The 'muscular sense' was now fully evidenced.

Sherrington did try to get this message out. A few weeks before Christmas, 1900, he wrote to Ruffini:

> I have been earnestly engaged in organising teaching in 'Fisiologia' to school-masters & school mistresses of our state-schools. It appears to me very important that those who have the charge of children in school should know the facts about the normal working of the body & the senses.[9]

Yet, as far as most school curriculums are concerned, it still isn't a sense.

Muscle spindles are the stars of proprioception,[10] but there are a few important supporting actors. If you're resting this book on something, lift it up. Golgi tendon organs, which are packed into the junctions between your muscles and tendons, will register this and signal the increase in tension as you lift the load. (Ruffini did important work on these organs, too.)

The third main group of proprioceptors consists of mechano-receptors located inside and around your joint capsules – the seals around the edges of bones that make up a joint. This group includes Pacinian corpuscles, which as we know are also found in the skin. Here, their job is to report on even tiny vibrations through the skeleton. Also around your joints are Ruffini's own stretch-sensitive endings. When you rotate a joint, such as a knee or elbow, they will respond.

If stimulated, both Ruffini endings and Golgi tendon organs will send continuous signals to the brain. They are each both maximally stimulated at particular joint angles. Your brain can use the degree of stimulation of these receptors to update its map of the position of your limbs, while also incorporating constant updates from the other proprioceptors. Still, that early German idea that movement commands issued by the brain to muscles are important for our perceptions of our limbs moving wasn't entirely wrong. We already know that in generating perceptions of sounds, sights, and so on, our brains rely not just on sensory signals but on expectations. Exactly the same is true for proprioception.

A group of six brave Australian volunteers provided clear experimental support for this. In a study run at the Prince of Wales Medical Research Institute in Sydney, they agreed to have their right arms both paralysed and anaesthetised. They could neither move their hands, nor feel any sensations in them. But when they were asked to try flexing their hand at the wrist, they reported *feeling* that it was moving – and the greater the effort they put into this, the bigger the movements that they perceived.[11] The motor signals sent by their brains to the muscles in their hands didn't cause any actual movements, but, based on extensive past experience, their brains fully *expected* them to. These trusted expectations were duly fed into their brain's body maps, and the volunteers' own perceptions.

This body map is thought to be critical for so much that we do. Think back to Chapter 5 on touch and that wine glass in your kitchen cupboard. Before you can touch it, you have to get your arm and hand into the right position to grasp it. First, your brain plans the desired movement. Based on the current body map, and what's happened in the past, it predicts the outcome of instructions to specific muscles in your arm and hand. Then, as you start to reach for that glass, proprioceptive, visual and, later, touch signals guide and refine these instructions, so that – it is hoped – your fingers make delicate contact with the stem. If and when that happens, signals from both muscle spindles and Golgi tendon organs ensure that when you pick up that glass, the muscular power is matched to the mass . . . In other words, you pick it up gently, rather than smashing it, Hulk-like, up against the roof of the cupboard.

Through repeated attempts at picking up a wine glass, catching a ball, or playing the piano, our brains get better at coordinating our actions – at knowing what to expect and what to instruct – and we go from being slow, clumsy and deliberate to something like proficient. In 2019, researchers at the University of Pittsburgh published a study revealing the key changes in the brain as this happens (in this case, in the brains of rhesus monkeys): as the monkeys went from being complete novices at a physical task to experts, new neural patterns of activity began to appear.[12] Practice seems first to create and then strengthen these patterns.

We do of course often use vision to help us to coordinate our actions. If you think you could close your eyes, walk into your kitchen and smoothly pick out a wine glass from its cupboard, this suggests a certain amount of practice . . . Actually, if you were to close your eyes, you could probably get around your kitchen fairly well. If you suddenly lost proprioception, though, you'd have no chance. How do we know? In part, from people with first-hand experience of exactly what that's like.

In 1971, Ian Waterman was nineteen years old, working as a butcher on the island of Jersey. He became infected with a virus. His body's reaction was unusual – and catastrophic. Waterman's immune system turned on the nerve pathway that transmits information about touch and proprioception.[13] He was left able to sense other things, including pain and temperature, but proprioception and touch below the neck were gone.[14]

Since Waterman's motor neurons were working, he wasn't paralysed. But he couldn't stand or even sit up. Without any sense of the activity of his muscles and joints, he couldn't even pick up a cup. Lying in a hospital bed, he'd find his limbs floating off, hitting someone sitting to either side, or knocking objects from the bedside cabinet. His doctors told him that his condition was incurable and that he would spend the rest of his life in a wheelchair.

Waterman refused to accept this. Eventually, he learned to rely on vision to keep track of his limbs and to monitor his movements. But this takes an immense amount of effort. He has to scan the footpath, six to eight feet ahead, planning all the time for what's coming up, so he's aware of where he's going to place his feet. He also constantly checks his feet, always looking at the consequences of his physical actions. It took Waterman a year to learn to stand safely, and three years to learn to walk like this. And it has never got easier. With the permanent loss of proprioception, it seems that the use of motor signals for body mapping fades.

The loss of touch robbed Waterman of his instinctive knowledge of where he ended and the rest of the world – from a pavement to a fork – began. The loss of the nerve fibres that carry proprioceptive information took away his instinctive understanding of the location and movement of parts of his own body.

Waterman lost proprioception as an adult. The girl and the woman mentioned in the previous chapter who lacked Piezo2 grew up without it. They could walk almost normally – until, in tests, they were blindfolded. Then, they stumbled and fell. When the researchers moved their limbs for them, they couldn't sense what was happening. Having been born without proprioception, their brains had adapted dramatically, their sense of sight compensating massively, effectively taking over the job of monitoring the position of their limbs in space.[15]

These two individual patients also had skeletal deformities. They had scoliosis, a curvature of the spine, and their hips and feet were at unusual angles. This led the researchers to suspect that proprioception is important for typical skeletal development. In order to maintain your posture, your muscles are constantly sending signals about their shape, and receiving instructions in return. If that process is broken, you can't keep your posture straight, or hold your limbs in typical positions – and normal bone development goes awry.

However, if you do have typical proprioception, even if you have an unusual skeletal configuration, your adaptable brain can take this in its stride. Clear evidence of this comes from studies of people with extra body parts.

People who are born with an extra, sixth finger usually have it surgically removed at birth, as it's assumed that it can't be any use to them, and may draw unwanted attention from other children. However, in 2019, the first study of the physiology and sensorimotor mechanics of two people with an extra fully-formed finger between their thumb and index finger found significant advantages.[16]

Not only did these people regularly use their extra finger in daily life, but when tested on a video game that required coordinated key presses, they could achieve with one hand what a typically-fingered person could only do with two. This work shows that our nervous systems and brains are perfectly capable of adopting and coordinating extra digits (and what else – entire limbs?) It also implies that firstly, setting aside the psychological impact of looking different, extra fingers should not be removed and secondly, an extra, bionic finger might be useful for the rest of us.

While proprioceptive signals are essential for our intuitive understanding of our body's positioning, they may also have follow-on impacts on our psychological state.

Perhaps the best-known proponent of the idea that your posture affects how you feel is the Harvard social psychologist Amy Cuddy. Cuddy's 2012 TED talk on 'power posing' has been viewed more than eighteen million times.[17] A power pose is an open, expansive stance. You might have your legs widely planted, flat on the ground and your hands on your hips. In contrast, head down, shoulders slumped, arms crossed over the body, is a defeat pose if ever there was one.

There are certainly psychologists who, based on their own research and reviews of the literature, don't believe that the brain's interpretation of the patterns of proprioceptive signals generated by power posing have all of the effects that have been claimed.[18] Cuddy herself has relaxed her own metaphorical stance on this. But there is still evidence that a power pose does lead people to *feel* more powerful. In theory, this could affect the result of a meeting or a negotiation.

Think now of the New Zealand All Blacks and the women's team, the Black Ferns, and their hakas, which they perform before a rugby Test match. The haka originated as a way of preparing for war. A rugby player isn't highly likely to die if he mis-judges the position of his arms and legs in space; a warrior going into battle almost certainly is. It's easy to imagine how powerful, rhythmic, repetitive movements, which stimulate proprioceptors, might prime us for better using our bodies. A haka can function, then, as a proprioceptive warm-up. But there may be something about those low, solid, grounded, confrontational movements that makes the individual feel psychologically stronger, too. The lyrics of a haka are deep in meaning, and it's impossible to separate the effects of the words and the actions. But for Te Kura Ngata-Aerengamate, who jointly led her team's haka during the 2017 Women's Rugby World Cup, the effect is potent. 'It feels like a rumbling in your heart,' she told a reporter at the time. 'You get into this zone of extra energy, like you are on the next level.'[19]

For most of us, proprioception is usually a gentle sense. Unless we're straining our muscles or really loading our limbs, it doesn't yell at our consciousness, requiring attention, in the way that vision does. As well as being less demanding than vision, it's also less precise. This means that if proprioception is saying one thing and visual information is saying another, your brain will tend to go with its most trusted messenger, vision, and overrule your proprioceptors. Limb location's mechano-sensory sister, touch, is not as favoured, either. It, too, can be shouted down by vision. And these dynamics can make for some very odd illusions.

Can you get hold of a table-top mirror? If so, you can try a body-related illusion for yourself. Place one hand on the table in front of the mirror, palm up, and the other behind the mirror, so you can't see it, palm down. Now open and close your hands.

Within about a minute, you should feel as though the hand behind the mirror has suddenly flipped to match the reflection of your other hand. (Jared Medina at the University of Delaware describes what happened when he ran this experiment in the lab: 'All of a sudden, you'd hear a little laugh of surprise when people experienced this neat sensation of feeling like their hand flipped, even though it did not move.')[20]

If you tried it, and it worked for you, touch and proprioceptors would have been telling you that your hidden hand was maintaining its position. But your eyes were suggesting otherwise. Since vision is usually an accurate source of information about how your body parts are oriented, to resolve the sensory conflict, your brain plumped for vision.

This illusion belongs to a bigger family of 'rubber hand', 'rubber body' and even 'rubber tongue' illusions, all of which demonstrate just how vulnerable our perceptions of our own bodies are to manipulation.

A simple version of the rubber hand illusion, which was first described more than twenty years ago,[21] goes like this: a volunteer sits down with their forearms resting on a table. A curtain is used to conceal their right hand. A rubber hand is then placed in front of them, in line with the right shoulder, the stump covered by a cloth. The researcher uses soft brushes simultaneously to stroke the fingers of the real, hidden hand and also the fake hand, which the volunteer focuses on. Before long, most people will report that the rubber hand starts to feel as though it actually belongs to their own body, while their own, covered, hand feels less real. (I did try this recently and I didn't feel it; however, it was at an evening event, and a few glasses of wine had sent my heart rate into disarray. I was very conscious of a rushing of blood and a tingling in my real arm. The fake hand stood no chance.)

It has since been shown that people can come to experience an entire artificial body as their own, and if they are watching a mannequin being stroked on the chest, they can feel stroking on their own chest. Another experiment, dubbed 'Being Barbie'[22], has even found that it's possible to induce the illusion that you are the size of a thirty-centimetre-by-eighty-centimetre doll (this is achieved through synchronous touching of the participant and a doll), making everyday objects seem enormous.[23] For his part, Charles Spence has found that not only can synchronous stroking lead to a feeling that a fake, rubber tongue is one's own, but when lemon juice is applied to that rubber tongue, some people will even perceive a sour taste.[24]

Henrik Ehrsson, a cognitive neuroscientist and expert in these body-related illusions, reports that they work on most people (usually 70 to 80 per cent) but not all. It's not clear why some people don't succumb, but it could be because they are better at proprioception. Ehrsson himself says that while, like him, one of his two brothers experiences the rubber hand illusion vividly, the other doesn't. This brother also happens to be a proficient guitarist, used to relying on proprioceptive signals about what his hands are doing.[25]

For those who do fall for these illusions, there can be powerful biological effects. In people made to feel that a rubber hand is their own, there's a small but measurable drop in the skin temperature of their real hand,[26] indicating that blood flow has been reduced. What's more, their immune system, which takes its job of discriminating 'own-body' cells from 'not-self' cells extremely seriously, responds too, increasing levels of histamine in the real hand.[27] Raised levels of histamine are implicated in autoimmune disorders, in which the immune system turns on the body's own cells. The illusion is enough, it seems, to make a person's immune system start to disown a genuine body part.

However, although vision can dominate proprioception, signals from the limbs can also influence what we see. Again, this is because expectations, not just sensory signals, feed into our perceptions. Of all the studies that demonstrate this, there's one that for me really stands out.

It's another that you can in theory try for yourself. You'll need something that works as a complete blindfold but allows you to keep your eyes open. A VR headset would do.

Sit with the blindfold on, raise one arm, bent at the elbow, and slowly wave your hand back and forth in front of your face, from one side of your head to the other. Do you *see* anything? If you do, do you see motion? If so, does the moving object have a form?

In the original study, led by neuroscientist Duje Tadin, when an experimenter waved his hand in front of blindfolded participants, none reported seeing anything. This was obviously no surprise. But when they were asked to wave their own hand instead, half reported visual sensations. Many of this group talked about seeing motion, often describing a moving shadow or a darkening. These sensations tended to be stronger when their hand was in the peripheral, outer area of their visual field, rather than in the centre.[28]

Were they really 'seeing' something? Or were they just reporting that they were?

When you track the motion of an actual moving object, your eyes make movements that are subtly different to those made when you track an imaginary object. Using tiny cameras, Tadin was able to confirm that when the blindfolded, hand-waving people in his study claimed to be following something that they could see was moving, in a very real sense, they were. Their brains and eye muscles were behaving as though an actual object was genuinely there. These perceptions were stronger in the periphery presumably because, as Anil Seth has found, we rely less on raw retinal data in constructing our representations of this portion of our visual field.

This study shows that, for many people, proprioceptive signals associated with a particular movement are enough to generate the kinds of visual perceptions that typically accompany that movement. Tadin observes that this phenomenon might help to explain ghost sightings, which of course are often reported at night. A ghost-hunter flailing about in the darkness might see a strange flash of movement and give it a spooky interpretation, when it's just their brain hallucinating the motion of their own arms.

It's clear that proprioception is critical to our everyday life. But can you get better at it? And, if you can, what are the benefits?

When it comes to individual differences in proprioceptive sensitivity, much less is known than for any of the other senses discussed so far. Given that there is a spectrum of touch sensitivity, though, it seems likely that the same is true for proprioception. This could help to explain cases of extraordinary hand-eye coordination, and sporting prowess, as well as the clumsier end of the spectrum.

Some people do seem to be preternaturally able. Not long ago, my husband took a friend, a former Premier League footballer who had never played golf before, out on a golf course. This man's performance was 'incredible'. He was hitting balls like someone who'd been playing for years. He must, I'm sure, have the 'right' genes for proprioception.

I ask Gary Lewin about this. 'I don't know of any very good way of quantifying how good proprioception is directly at the receptor level,' Lewin says. 'But I think that extremely agile and accurate athletes almost certainly have an incredibly good proprioceptive system. The motor movements you need to play tennis or another very fast sport really require instantaneous feedback from your muscles and tendons, for you to know where they are at any time. It's no coincidence that you can't train everybody to become a top athlete. It has to be something to do with the sensory system that you are born with.'

Still, as with all of our senses, practice at using it is crucial for its development. And there is evidence that this starts in the womb. A foetus's wriggling and limb-flailing may look random, but in the later stages of pregnancy, when a foetus's hand moves towards its mouth, the mouth will often open first, in preparation for sucking on it. Without proprioception, this wouldn't happen.[29]

Clearly though, babies are extremely uncoordinated. It takes a lot of practice for a child to learn to bring a cup to its mouth without spilling its contents first, or to construct a tower of building blocks, never mind to climb a tree. Some researchers worry that the increasing use of screens by young children is a threat to this sense, as well as to touch. Kids who spend too much time indoors, not running around, not swinging on trees, not broadly challenging their bodies, may become excellent manipulators of a PlayStation controller, but put them out in the real world, and they'll struggle.

And what about adults? Our ancestors were active hunters and gatherers. Now many of us spend long periods sitting at desks, on sofas and in cars. For older people, in particular, physical instability is a big problem. If you don't know exactly where your feet are, and you can't sense your bodily alignment properly, you are more likely to fall.

The good news is that proprioception, like our other senses, can be trained.

Northern Ballet, which is based in Leeds, is preparing to go on tour. Packing cases of props – quills and brass candlesticks for *The Three Musketeers*; rubber roast chickens and croissants for *The Nutcracker* – jostle for space in the corridors alongside crates of satin pointe shoes. Many of the shoes have been modified by their owners, the edges of the rigid shanks of the soles filed down with Stanley knives, thick stitches sewn in to reinforce the tips.

A pair of pointe shoes is the polar opposite of a pair of trainers. When flat on the ground, they're hard and unstable, and difficult to balance on. But a ballet dancer must often spend long periods *en pointe*, her entire body weight up on the tips of the shoes – or even just one. To perform a scene with grace, and not to wobble, a dancer must possess superb strength and stamina, and an incredible sense of proprioception.

The ballet mistress at Northern Ballet is Yoko Ichino. In the course of her long dancing career, Ichino has danced with any number of ballet legends, including Mikhail Baryshnikov ('They were looking for a small dancer, because he is small, though he doesn't think so!' she laughs), and Rudolf Nureyev, with whom she danced in Sir Peter Wright's renowned production of *The Sleeping Beauty*. The 'Rose Adagio' in Sleeping Beauty has a reputation for being particularly challenging when it comes to balance. 'It's quite long and it's almost all on one leg, which gets tired and cramped. There's a *lot* of physical control,' Ichino says, shaking her head.

Back in the 1980s, Ichino devised a training technique to improve her own physical control. Now, she teaches it to young students at the Northern Ballet School, as well as to the company dancers. An important element is learning to be intensely aware of your anatomical alignment. (If your weighty head is directly above your spinal column, it requires less effort to keep your awkwardly inverted, segmented pendulum balanced.) Proprioception plays a fundamental role in this. And to train proprioception – to support not only perfect alignment, but control on the stage – Ichino took inspiration from the experiences of Alicia Alonso, a Cuban *prima ballerina assoluta,* the title that is given only to truly exceptional female dancers.

Born in 1921, Alonso started studying ballet as a child in Havana. But at twenty, her vision deteriorated, and she had the first of a series of surgeries for a detached retina. Alonso's recovery included long periods of bed-rest, with her eyes kept covered. While in this state, and with the help of her ballet-dancer husband, Alonso learned how to play the title role from *Giselle* by *imagining* the moves. Ultimately, she was left partially blind and without peripheral vision. But remarkably, she went on to give highly acclaimed performances – including the role of Giselle – for the American Ballet Theatre. 'We appeared on the same programme at many galas and festivals together,' says Ichino. 'Early on, my question was, "How do you do that when you *can't* really see?" So I used to try to walk around my flat with my eyes closed, and see if I had memorised where everything was and how much space there was. And I decided we use our eyes too much. We need to use all our other senses as well, but because our eyes are open all the time, we never develop them. So I put that into my own training and then the training work. And I find that people can learn faster.'

Ichino leads the way to Studio 4, and her 10am proprioception class. Already assembled are several dancers who have recently joined the company, as well as some graduate students. It's a bright, light, high-ceilinged room, with a barre running along three sides. On the long fourth side are floor-to-ceiling black curtains. They are closed, to conceal the mirrors. For this class, Ichino doesn't want the dancers to be able to see what they're doing. (Even the younger students get to see their reflections only very rarely. 'I want them to *feel* what they are doing,' she explains. This is very different to other ballet schools, where mirrors are the norm.)

Wearing a long-sleeved purple top and grey tracksuit bottoms, Ichino stands with her back to the curtains. When the dancers are in position, and the pianist has readied his music, she begins to issue instructions in a shorthand of attenuated gestures and oblique verbal commands. Though it's bewildering to an outsider, the dancers all seem to understand. The first part, I do catch: 'We're going to go right into *pliés*. You will be doing this with your eyes *closed*.'

I watch Adam Ashcroft, a tall twenty-two-year old who trained at the Royal Ballet School and who recently joined the company from the Estonian National Ballet. As he stands and bends his knees, feet turned out, then straightens his legs again, signals from muscle spindles, joint proprioceptors and Golgi tendon organs race to his brain. With his eyes closed, the only way Ashcroft *knows* that his legs are moving into a *plié* is because of his sense of proprioception.

As the dancers glide into other moves and positions, Ichino monitors them closely, walking between them, smiling but frequently making corrections. 'Where's your belly button? *In!*' 'Diagonal body!' 'Eyes *closed*!'

When she then instructs them to pick up plywood stools, to hold above their heads ('*eyes closed*'), 'arms straight', while they perform foot-flipping *coupés,* not only new proprioceptive, but new touch signals start streaming into their brains. The load on Ashcroft's muscles means Golgi tendon organs as well as muscle spindles and other proprioceptors in his arms are stimulated. His anatomical alignment and his balance are challenged by the weight of the stool. The straighter he can hold it above his head, the easier it will be to keep his balance. With his eyes closed, his conscious attention is focused on the information streaming in from his muscles and his skin. It's not on how his body looks, or how the rest of the world looks, relative to him, but how he feels from within.

When the eighty-minute-long class has finished, I ask Ashcroft what he thinks about this approach to training. It's very different to what he's experienced before, he tells me. 'If you close your eyes while dancing, it makes the weaknesses glaringly obvious. You can easily see the kink in the chain of your line of balance.' After just a few weeks of training with Ichino, he reports already being much more stable.

Moving to music is one obvious way to train proprioception, to improve your stability and control. Of course, some cultures value such practice and achievement more than others. When making music to dance to, the Tiv of Nigeria use four drums, each playing a different beat, one for each part of the body.[30] This is proprioceptive training on steroids.

For a non-dancer, though, there are many kinds of ways to get better. Just picture a child's idea of great fun – away from a screen. A typical playground, with ladders, monkey bars, an obstacle course and maybe a rock-climbing wall, *is* a proprioceptive-training zone. In fact, it's hard not to think that the reason children love playgrounds is because the need to develop good proprioception is so vital to our survival that we evolved to feel a sense of reward from engaging in these training activities.

Climbing trees, walking along balance beams, navigating obstacles, crossing stepping stones (which you can simulate at home, using small mats placed on the floor) – all of these are proprioceptively demanding. Researchers at the University of North Florida who tried all these activities with adult volunteers found improvements not only in physical coordination but also in working memory (your capacity for holding and manipulating information in your mind).[31]

Specific programmes of proprioceptive training exercises have also been devised for older people.[32] Recently, a team of rehab and sports physicians in Shanghai, China, compared the effects of a sixteen-week programme of forty-five-minute, twice-weekly, sessions of either tai chi or a proprioceptive regimen on a group of healthy seventy- to eighty-four-year olds. Tai chi, an ancient Chinese practice, involves a series of smooth, steady, controlled body and head movements. These classes were led by an experienced tai chi teacher. The proprioceptive sessions were led by a physical therapist, who guided the elderly volunteers through a warm-up, followed by twenty minutes of static exercises, such as squats, then fifteen minutes of movement exercises (jogging sideways, walking backwards, or running in a zigzag line, for example), followed by a cool-down.

At the end of the study period, the team assessed the volunteers' ankle joint position-sense, and they found similar significant improvements for both groups. However, they noted, the tai chi group did report liking their training more.[33]

You might also borrow from Yoko Ichino's inspiration, and try sometimes to move around your house with your eyes shut. It would also be worth remembering her class, and, if you know yoga or Pilates moves, doing these with your eyes closed. My own Pilates teacher often gets the class to engage in moves with our eyes shut. It's revelatory. Turn off vision, and you are instantly so much more aware of those signals from your body. Blot out the Sun, and this gentle sense, Moon-like, can shine.

The bonus is that, in moving around with your eyes shut or climbing a tree, you'll challenge not only proprioception but another group of senses too. Our senses very rarely operate in isolation. And to stay balanced, we need signals not only from our muscles, but also from the vestibular system in the inner ear. When you think of the ear, you, like Aristotle, no doubt think of hearing. And yet your inner ear probably evolved initially not for hearing at all, but for these other, critical sensing tasks.[34]

7

Gravity and Whole-body Motion

How to be a whirling dervish (without falling over)

Imagine that you're sitting in the central section of a packed passenger jet, waiting for take-off. Vibrations from the jet engines are rippling through your seat, but you can feel that the plane isn't taxiing yet. Then, although you can't see through a window, you have the unmistakeable sensation of moving forwards. The sounds and the vibrations help to reveal the fact that the plane's now speeding along the runway, but that evidence is practically circumstantial, because you just *know* that you're crossing ground fast.

Suddenly, you become aware that the nose of the plane has lifted. The signals from your eyes have not changed. You are in precisely the same position relative to everyone around you and everything that you can see. But you now have the unshakeable conviction that you're on a steep ascent. How?

It is thanks to the same system that tells you, with no need for vision, when an underground tunnel starts to curve, when you're shooting upwards in a lift, or are upside down on a rollercoaster – or, for a ballet dancer, when their head is spinning in a pirouette or dropping in a melting *fondu*.

This system provides precise information about the orientation and movement of your head in space at all times. No, aeroplanes, lifts and rollercoasters were not around when humans evolved. But hills and holes, uneven ground, and darkness were. Without a vestibular system, our ancestors would have struggled to keep upright. At night, they wouldn't even have known where 'up' was.

Right beside the cochlea, which enables hearing, are three semi-circular canals and two 'otolithic organs', which together make up your vestibular system.[1] Both the cochlea and the vestibular system use the same fundamental sensory apparatus: hair cells housed in liquid, which convert mechanical signals into electrical signals for the brain. While sensory hair cells in the cochlea react to incoming pressure waves, those in the vestibular system are triggered by you moving your head – or by the simple force of gravity.[2]

Let's take a closer look first at the semicircular canals. These are interconnecting tubes, each arranged at ninety degrees to the other. When you nod or shake your head, or tilt it – or something, like an aeroplane or a rollercoaster car, tilts it – the liquid inside sloshes around, bending sensory hairs, which fire off signals. If your head isn't moving, the liquid pools, and this is also registered by your brain. When your head is still, nerves leading from sensory hair cells within the semicircular canals fire at about ninety times per second. Move your head, and this rate changes in proportion to the acceleration of the liquid within any given canal. This gives your brain very clear information about not only how your head is oriented, but how fast it's moving.

Between the semicircular canals and the cochlea are our two otoliths ('ear-stone' organs). They consist of sacs containing force-sensitive hair cells, whose tips are embedded in a gelatinous membrane packed with calcium carbonate crystals (these are the 'ear-stones').

In the saccule (from the Latin for 'little sac'), this membrane is vertical. This makes it sensitive to gravity, and also capable of detecting up/down movement. In the utricle (the diminutive of the Latin for 'leather bag'; 'uterus' has the same etymological origin), it is roughly horizontal. Whenever you walk along a street, or ride in a car, utricle signals will tell your brain that you're moving, but also how fast you are going.

The gravity-detecting saccule gives us an instinctive understanding of which way is 'up'. For even very primitive animals, this is critical information. In fact, there's evidence that a saccule-like receptor for detecting the direction of the pull of gravity existed in early life forms.[3] Invertebrates, such as mussels and sea snails, and even plants have something very like it. Comb jellies, which may have been the first multi-cellular organism to evolve, also have a simple, but highly effective, version.

Squid and octopuses have slightly more complex systems, which respond to sideways as well as up-and-down movement. And though these species can't 'hear', their statocysts are certainly sensitive to the vibrations of low-frequency sounds. This discovery was made in the wake of two sudden spikes in the numbers of dead giant squid washing ashore on the west coast of Spain in 2001 and 2008. Though the autopsies had failed to find an obvious cause of death, bioacoustics researchers in Barcelona noted that seismic surveys of the ocean floor in the region had been conducted at both of those times. These surveys use intense pulses of low-frequency sounds. The team found that such sounds can badly damage the statocysts of not only squid but also octopus and cuttlefish. The surveys would have left the giant squid unable to orientate, the researchers concluded. They think that they most probably died after they floated to the surface, and were unable to eat.[4]

This work adds to other research showing that animals without auditory systems can be affected by noise vibrations. But it also helps to build a picture of inner-ear evolution that starts with a gravity detector, and then vestibular vibration-sensing, with hearing coming later.

As is the case with limb-location, these vestibular senses have been known in scientific circles for an awfully long time. In 1889, a good ten years before Sherrington set out to try to educate the nation's school-masters and mistresses on the senses, Christine Ladd-Franklin, an American psychologist and mathematician, published an article in the journal *Science* under the headline 'An Unknown Organ of Sense'. She wrote: 'it is probably unknown to a great many of the laity that within a few years past a new organ of sense has been discovered, the existence of which had not before been so much as suspected . . .'[5]

Experiments had now confirmed, Ladd-Franklin went on to explain, that the semicircular canals, which had been discovered in the inner ear back in 1824, had nothing to do with hearing but were the organ for sensations 'whether conscious or not' that enable us to determine the direction and degree of rotations of the head.

Along with limb-location, the vestibular senses are vital for what's sometimes called our 'sense' of balance. After all, to stay upright, you need to understand where up is, and when you are tilting, and by how much. When the dancers in Yoko Ichino's class effortlessly folded their torsos down over their thighs, or began to leap across the floor (they were allowed to have their eyes open at this point), proprioception gave them the essential sense of what their bodies were doing, but their vestibular senses were critical for keeping them on their feet, and also registering their movement through space and their head position. An exquisite sensitivity to these signals can even allow for some truly jaw-dropping accomplishments.

The French high-wire artist Philippe Petit has performed death-defying cable walks all over the world. Perhaps the most famous of all was an unsanctioned walk that he undertook in 1974. Early in the morning of 6 August, Petit and his crew rode to the 110th floor of one of the two World Trade Center towers in New York City. They used a bow and arrow to fire a fishing line forty-two metres across to the twin tower. Attached to that line was a steel cable, which collaborators secured in place, 410 metres above the ground. Just after seven o'clock that morning, Petit picked up his balancing pole and stepped out onto the cable. To the astonishment of people below – and the consternation of the New York Police Department – he spent forty-five minutes walking, dancing and even lying down on the wire.

To prepare for walks like this, Petit would practise in the worst conditions. He might walk at dusk; with a balancing pole that was too heavy and so liable to tip him up; with friends hitting the cable; in high winds. By challenging his balance to the utmost, he hoped that a performance walk on a calm day would be a relative breeze.[6]

Petit embarked on the Twin Towers walk with his eyes literally open. Sight helps us to stay oriented in our environment, and so is important for balance. (You can check just how important by standing up now and lifting one foot off the ground. Can you hold this position? Now shut your eyes. If you could easily stand one-legged before, once deprived of vision, I bet you started to wobble.) Petit, however, has even performed cable walks while blindfolded.

One of the most remarkable blindfolded high-wire performances ever has to belong, however, to the American aerial performer Nik Wallenda. In 2014, in front of journalists who had signed waivers relinquishing the right to claim emotional distress if the worst happened, Wallenda wore a blindfold while he walked a wire that stretched 543 feet above the ground between the two Marina City skyscrapers in Chicago.[7] For around a minute, Wallenda's life was in the hands of his vestibular and limb-location senses – senses that many of us don't even know that we have.

Normally, the brain uses visual as well as vestibular information to keep you on task. This applies to your eyes, as well as the rest of your body. When vestibular signals indicate head motion, they also trigger motor commands that create compensatory eye movements, allowing whatever you're looking at to stay in the centre of your visual field. This is known as the vestibulo-ocular reflex, and you can test it now. Try holding a finger up in front of your eyes and looking at it while you twist your head from side to side. If you turn your head to the left, your eyes will swivel to the right, and vice versa. Your finger should remain in focus.

When signals from the vestibular system and the eyes don't match up, however, you can be left feeling queasy. At least, this is the leading explanation for motion sickness, which has fascinated doctors all the way back to Hippocrates, who noted that 'sailing on the sea shows that motion disorders the body'.[8]

Back in 1968, a team from the US Naval Aerospace Medical Institute in Florida studied twenty healthy people and ten patients with serious vestibular defects in both ears while they were on a boat in the North Atlantic during a storm. The researchers dispassionately observed that most developed 'a feeling of fear'. However, while all members of the healthy group developed typical symptoms of motion sickness, including vomiting, none of the patient group did.[9] This showed that vestibular signalling was critical for motion sickness to occur.

If you're sitting in a windowless cabin in a rocking boat, and you have a healthy vestibular system, it's telling your brain in no uncertain terms that you're being thrown about. But because everything is moving with you, your eyes are maintaining that you're stationary. This conflict is thought to be the source of the problem. Why should it make you throw up? It may be because various poisons – including alcohol – mess with visual and vestibular signals, and if you've swallowed poison it's obviously a good idea to vomit it out.

But why on any given ferry trip will you observe some people groaning, ashen in a corner, while others are happily chatting away, unaffected? No one really knows. But another study in which motion sickness was experimentally induced in fish (yes, really) suggests that, for healthy people, variations in the mass of the 'ear-stones' in the otoliths in the right versus the left ear could play a role in determining who has to rush for a sick bag, and who doesn't.[10]

The only known way to prevent or overcome motion sickness is to try to resolve the sensory conflict, so that signals from both the eyes and vestibular system are in agreement with each other. If you're in a windowless cabin on a boat, that means going up on deck, so you can see the waves as well as feel them. If you're in a car, it means putting away your book or phone and focusing on the world outside your window. These are simple strategies for working with your vestibular system. But there are more dramatic ways to harness it – with potentially far-reaching results.

We are all familiar with the soporific effect of gentle rocking. This, too, has been traced to the vestibular system. In 2019, a Swiss team showed that adults who slept in a bed that rocked back and forth every four seconds, moving a total of ten and a half centimetres, slept more soundly, and did better on memory tests. A companion study on mice found something very similar in healthy animals – but not in a group that had been bred to lack functional otolithic organs. The researchers aren't sure exactly what's going on, but they think that in causing rhythmic otolithic signals to be despatched, rhythmic rocking may exert a kind of synchronising action in the brain, aiding deeper sleep.[11]

Later that same year, a separate team reported a study that found vestibular stimulation could even help with anxiety. Student volunteers had electrodes placed behind their ears, to stimulate the nerves that carry information from the otoliths and semicircular canals to the brain. After three thirty-eight-minute sessions, anxiety levels in this group dropped by a quarter. (No such effects were found in a group that received sham treatments.) Some reported feeling that they were tilting or spinning during the treatment, but none felt motion sickness.[12]

The electrical currents used in this study were very weak, and they caused only mild sensations. If you were to start actually spinning around in a circle, thanks to the vestibulo-ocular reflex your eyes would move in the opposite direction to the motion of your head. But, of course, they can't swivel full circle. So when they've gone as far as they can, they'll jump back to a new starting position. And if you keep spinning, these repetitive eye movements will make you feel dizzy.

In a ballet move such as a *fouetté* – a whipping movement of a raised leg, usually accompanying a pirouette, which requires spinning – it's standard for ballet teachers to encourage students to keep their head still, their gaze fixed on a single point until the last moment, then to whip the head around quickly, and to repeat, to help prevent dizziness. 'This is *not* what I teach,' Yoko Ichino tells me. 'The position of your eyes is not so important because it should be your *body* feeling where you are.' After she has performed twenty *fouettés*, perhaps she won't see clearly, she says, but she insists that she won't feel dizzy. 'Because my *body* knows my orientation, not my eyes.'

However, it's certainly possible that a career involving an awful lot of spinning has caused Ichino's brain to change, to cope. Evidence for this comes in part from fascinating studies of members of an Islamic Sufi order founded in Konya, Turkey in 1273 by Jalaluddin Rumi, a Muslim teacher and scholar. Rumi used to whirl while meditating and composing poetry. After his death, his followers adopted whirling as a form of meditation. One of their Sema ceremonies, which can last up to an hour, involves spinning in an anti-clockwise direction while also rotating around other participants. Members of the order are known as the Mevleviye – or the 'Whirling Dervishes'.

A 'Semazen' taking part in a Sema ceremony rotates to the left, using their right foot to drive their body around, and takes a circular path. They are meant to keep their bodies relaxed and their eyes open but unfocused, so that their vision blurs. To the sound of at least one singer, a flute-player, a kettle-drummer and a cymbal player, they spin, and spin, and spin . . .[13]

Training to join the order traditionally takes 1,001 days. This training includes the practice needed to be able to spin for so long without falling over or even feeling dizzy, and there's evidence that this changes not only the way the brain processes vestibular signals, but the brain itself.

In 2017, ten members of the order who had taken part in an average of two whirling sessions per week for ten and a half years agreed to have their brains scanned. The results showed that areas of their cortex involved in motion perception were significantly thinner than is typical. The researchers suspect that extensive spinning has caused brain changes that reduce the perception of movement during whirling – allowing them to stay upright and in control during a ceremony.[14]

The Sema ceremony is, of course, religious. One of the intentions is to loosen the 'material self'. The whirling takes place on a round floorspace, which symbolises the universe. The Semazen's waistcoat, which he removes when he begins the whirling ritual, symbolises his ego. His long, cylindrical headpiece is shaped like a traditional tombstone, to symbolise the tombstone of his ego. His flowing white overcoat is the shroud of his ego. Both arms are raised, his right hand open-palmed, to indicate reception from God. As he spins, he dissolves into divine existence and makes contact with eternity.

There's a very deep spiritual meaning to the ceremony. There's also evidence that unusual stimulation of the vestibular system, caused by spinning, could be directly implicated in the whirler's spiritual experiences.

Though the Mevlevi Order has taken religious spinning dance to an extreme, wild dances, involving rapid movements of the head and spinning, are found in other religious traditions, such as voodooism. Voodoo ceremonies involve various styles of dances, but it's during the wilder dances that trances are most likely to occur. For believers in voodoo, a dancer who falls into a trance has allowed their body to be possessed by a spirit. In terms of what they actually report, there can be radical changes in perceptions, including hallucinations.

Szuzsa Parrag, a German-Hungarian woman who for the past eighteen years has been studying voodoo dance under the tutelage of a priest in Haiti, explains it this way: 'Sometimes, I see cosmic spirals, in different colours. All the physical boundaries of the world disappear; I feel light, and deeply connected to the people around me, and the universe, and from there I can go anywhere. I feel very calm, and relieved from my fears. There's an inner voice that tells me "it's all good, and there is a continuous cycle of infinity, of life.'"

Reports of a sense of physical lightness, and being 'outside' the body, are something French neuroscientist Christophe Lopez is familiar with – from the stories of his patients with disorders of the vestibular system. Sometimes, these patients have infections of the inner ear, or inflammation of the vestibular nerve, though the precise cause is not always clear. They generally report severe dizziness. But, Lopez notes, another very common symptom is a feeling of lightness, which is usually a sign of dysfunction of the otolithic system.

In 2018, Lopez and Maya Elzière, of the Centre des vertiges at the Hôpital Européen, in Marseille, reported a study of the experiences of 210 patients presenting with severe dizziness, plus another 210 healthy people for comparison. Almost three times as many of the patients said they had had at least one out-of-body experience (14 per cent, compared with 5 per cent among the healthy group), and the vast majority of this patient group reported experiencing more than one.[15]

Most of these patients talked about sensations that Lopez and Elzière could trace to vestibular problems, such as feelings of lightness and of elevating above their own body. One talked about a 'sensation of being attracted by a spiral, like in a tunnel'. Another described a 'sensation of entering my own body, from the top'.

Out-of-body experiences have been called (by more spiritually-inclined commentators) instances of 'soul travel' or 'spirit walking'. This work suggests that disruptions to normal signalling from our gravity-sensing system can interfere with our fundamental feeling of being located 'down' and 'in' our own body.

This idea gets support from other studies, too. Though there is no 'vestibular cortex', as there is a visual cortex, say, stimulation of the temporoparietal junction – a part of the cortex known to receive vestibular information and to incorporate it with other sensory signals, including limb-location and touch data – can provoke out-of-body experiences. Such experiences may reflect a failure of the brain to integrate touch and vestibular information, argue Swiss-based neurologist Olaf Blanke and colleagues, in a paper published in *Nature*.[16]

Far from being 'just' an organ needed for balance, then, the vestibular system also plays a crucial role in our sense of bodily awareness. Perhaps in the context of a group religious ritual, such as a voodoo or a Mevlevi Sema ceremony, feelings of lightness caused by disturbances to vestibular signalling are more likely to be interpreted as a transcendent event than such feelings experienced by someone spinning around in a Sainsbury's car park, for example.

Most of us don't whirl, or spin in wild dances. But one of humanity's favourite drugs interferes with vestibular functioning.

Rumi's writings include references to wine. So do those of Omar Khayyam, a famous Sufi poet. It's been argued by some scholars that these references were all metaphors for spiritual intoxication. But wine does have vestibular effects that aren't entirely removed from those caused by spinning.

What's the classic test for drunkenness? Asking someone to walk in a straight line. What do really drunk people do? Stumble and fall over. What's one of the worst symptoms of being drunk? Feeling as though the room is spinning. All of these implicate the vestibular system.

In one study, ten healthy young men agreed to drink one and a half millilitres of whisky per kilogram of body weight within five minutes, all in the interests of science. The average British man weighs about eighty-three kilograms, which would mean about 124 millilitres of whisky – around four shots. This 'moderate' (in the researchers' words) amount of whisky clearly affected their ability to keep themselves fully upright, without swaying. The team's conclusion after various tests was that the vestibular system had been compromised. They think that too much alcohol changes the concentration of fluid within the semicircular canals, which leads to conflicting signals from the eyes and the inner ears about what the head is doing – and this causes the horrible spinning sensation.[17]

That level of alcohol intoxication is not pleasant. But ask people why they drink alcohol, and one of the reasons they'll give is that it blunts reality, and takes you 'out of yourself'. Alcohol has all kinds of impacts on the brain and how we feel. But perhaps one of the psychological impacts that at least some of us crave is partly driven by changes in those tiny canals, deep inside our ear.

*

Though the patients seen by Elzière and Lopez had serious vestibular problems, there is clear variation in the functioning of these
senses even among healthy, young people. And these differences can
have life-changing effects.

One way of testing vestibular sensitivity is to seat people in the
dark on a chair that can tilt, and then determine the smallest motion
that they can sense. A recent Harvard-led study using this technique
found that people who were more sensitive also did better on a
piloting task in which they had to use a joystick to keep themselves
upright.[18] What's more, the team found that this task was harder in
low gravity than in high gravity. This kind of vestibular assessment
might then be useful for screening astronauts before crew assignments.
If you're picking the pilot to land a craft on Mars, for example,
you'll want someone with stand-out vestibular sensitivity scores.

Vestibular sensitivity varies in young people, but for all of us it
declines with age. From the age of forty, vestibular system thresholds
(measured by something like the tilting-chair task) double every ten
years.[19] The older we get, the duller our vestibular responses become.[20]

The team at the Massachusetts Eye and Ear Infirmary behind this
finding also noted that people with poorer vestibular sensitivity do
worse on tests of balance, and that people who do badly on these
tests have a much higher risk of falls – something that can be especially dangerous in older people. In fact, based on a series of studies
that they have conducted, this team estimates that vestibular dysfunction could be responsible for as many as 152,000 deaths in the US
alone every year.[21] If they're right, this would put it just below heart
disease and cancer as the third-leading cause of death among
Americans.[22]

Other research supports this bleak picture. A separate team at the
Johns Hopkins University School of Medicine conducted a three-
year vestibular study of more than 5,000 men and women aged over
forty. They gave balance tests and threshold exams, looking at who
had noticeable vestibular dysfunction, and who didn't, and what the
early signs and symptoms might be.[23]

The team found that the risk of vestibular dysfunction increased
steadily with age, and also diabetes. Eighty-five per cent of people
aged over eighty had vestibular problems – twenty-three times more

than for people in their forties. (People with diabetes were at a 70 per cent higher risk, probably because of damage caused by high blood-sugar levels to the hair cells and small blood vessels within the vestibular system.) Strikingly, about a third of people in this study already had some vestibular problems at the start, but weren't aware of them – and they were three times more likely to suffer a fall than those who started out with healthy vestibular responses. Volunteers in the study who were already aware that they had balance problems had a twelve times higher risk of falling.

Many middle-aged and older people with vestibular dysfunction go undiagnosed, it seems, their difficulties often only being noted after an accident, such as a fall. This work shows that balance-testing needs to be a part of basic primary care, argues study investigator Lloyd B. Minor, director of otolaryngology – head and neck surgery – at the university.

Still, the statistics tell only so much of the story. For actual older people who know they have balance problems, the *fear* of falling alone can have all kinds of harmful impacts, such as refusing invitations to leave the home. One eighty-eight-year-old female participant in the British study that asked older people to talk about their sensory problems and their lives made this comment: 'I wouldn't want to be a bother to my family, if I broke my arm or whatever. So I've got to be sensible. Even though I'd miss going out, I say . . . . "No, thank you."'

There are various ways that vestibular dysfunction could make falling more likely. Clearly, people who aren't quite sure how their body is oriented would be more likely to sway and stumble. Vestibular problems can cause general dizziness, too. And a Harvard team, working with NASA scientists, has shown that stimulation of the otoliths has a direct effect on blood flow to the head.[24] Degradations of the utricle and the saccule could cause temporary dips in blood supply to the brain, leaving you light-headed and unstable.

For people with clear vestibular problems, there are exercises that can help. A doctor is best placed to advise about these, but there's also plenty of information online (just search on 'vestibular rehabilitation'). For the rest of us, many of the exercises that challenge limb location-sensing train the vestibular senses too. Dynamic movements, like those involved in climbing or Pilates or tai chi, which require movements of the head, and practice at keeping your balance, will all help.

If you have kids, you might even raid their toys. One of my own boys' favourites for summer afternoons is a basketball-sized ball with a broad rim around the equator. (It looks a bit like a plastic Saturn.) You have to stand on the rim and keep your balance. They'll spend ages doing this. And it's excellent vestibular training. In fact, just as most kids love proprioceptively-demanding activities, they adore vestibular stimulation – and again you can find this for free in most playgrounds. What is a slide if not a saccule-stimulator? What's a roundabout if not *the* perfect way to send your semicircular canals, not to mention your otoliths, into overdrive?

In fact, think about what kids seem biologically programmed to enjoy – from climbing things to building towers, playing with sand to finger-painting. If you were to ask a sensory expert to design a programme to develop all their many senses, this is exactly the type of stuff they'd come up with. Kids are driven to use and improve their senses, because they are so invaluable for their ability to thrive.

Adults can't reasonably play in playgrounds, at least not without a child. But we can surf, drive racing cars, ski, ride on rollercoasters and visit water parks to whizz down the slides and rides – and get our vestibular highs in these ways.

You don't *have* to know about the vestibular senses to enjoy them, and to use them. But understanding them gives us a deeper insight into what it is to be human. From a simple sac that allowed an ancient sea animal to orient itself in the ocean, the vestibular system has come a long way – as has our understanding of what it does for us. In fact, of the many Cinderella senses, the vestibular senses are perhaps the most under-appreciated of all.

The vestibular senses are clearly important for keeping us safe, and so critical to our fundamental drive to survive and thrive. But there are other senses that are far more essential. In fact, though they languish in obscurity, if I could somehow wipe them from your body right now, you would have, at most, a few minutes to live . . . .

8

Inner-sensing

Go deep-sea diving on a single breath

What's the scariest thing that's ever happened to you?

I can think of a few contenders. Climbing, as a kid, over a gate onto private land, thinking that this had to be a risk-free transgression – only for an Alsatian guard dog to race at me, barking. Walking home from the tube late at night, feeling sure that someone was following me. Finding myself driving through a field planted with land mines towards an Israeli checkpoint, gun shots sounding in the darkness.

Think back to how you felt at that time. Maybe, at the memory, you're even starting to feel an echo of it now . . . I know that I am.

That first pang of alarm? That's caused by nervous-system signals from the hypothalamus in the brain racing to the adrenal glands, causing a rush of adrenaline. As that adrenaline circulates in the blood, it triggers all kinds of changes designed to enable 'flight or fight', a phrase that we owe to the great Harvard physiologist Walter Cannon. Your heart rate rises, to boost vital blood flow to your muscles. Your breathing quickens, too, and small airways in your lungs open up. These changes allow more oxygen in, to meet the heart's greater demands. At the same time, blood is diverted away from non-urgent matters, such as digestion.[1]

While threats can certainly be physical, they can also be psychological. If a dog were suddenly to spring at you, or your boss were to dump a massive, urgent project on your desk, your brain and body would respond instantly in much the same way. If the threat doesn't immediately recede – if that dog doesn't turn out to in fact be next door's goofy Labrador, or your boss won't heed your plea for a longer deadline – a hormonal stress pathway kicks in. The hypothalamus communicates with the pituitary glands, which in turn send hormonal signals to the adrenal glands. This process initiates the release of glucocorticoids, including cortisol, which, among other things, increases blood levels of glucose, for energy. If and when the threat does pass, this process stops too, and other kinds of signals calm you down, switching you into 'rest and digest' mode.

But in all circumstances – whether you're sprinting away from that dog, in definite 'flight', or relaxing on a sun lounger, sipping a cocktail, in 'rest and digest', or just out at the supermarket – your brain needs to keep a close check on precisely what internal changes are happening. And it does this thanks to a network of specialised inner sensors.

Scientists have long recognised the vital necessity of this inner detection and regulation system. In 1878, the celebrated French physiologist Claude Bernard wrote:

> It is the fixity of the *milieu intérieur* which is the condition of free and independent life . . . all the vital mechanisms, however varied they may be, have only one object, that of preserving constant the conditions of life in the internal environment.[2]

Commenting on this in 1922, the English physiologist and medic John Scott Haldane wrote, 'No more pregnant sentence was ever framed by a physiologist.' Four years later, Walter Cannon gave the process of maintaining our *milieu intérieur* the name 'homeostasis' (from the Greek *homeo* – 'like' and the Latin *stasis* – standing still), by which it is still known today.[3]

For the body and brain to work well, and avoid damage and death, all kinds of factors must be kept within limits. If your heart rate drops too low, insufficient blood will reach the brain – and you will die. If your blood pressure soars too high, blood vessels will burst open – and you will die. If your lungs neglect to inflate, your oxygen supply will be cut off – and you get the picture. Fluctuations can be fine – even crucial, depending on the circumstances – but if any of these life-essentials bust their lower or upper boundaries, we're in real trouble.

By the time Cannon came up with the term 'homeostasis', physiologists had already worked hard to try to understand the limits of some of these variables. They knew that ordinarily, when there is a shift away from the average – if your blood pressure suddenly spikes, for example – something happens to rein it back in. Some of the processes involved in homeostasis were also understood. But in many cases, physiologists had no idea how a 'disturbed state' was sensed, or how it was brought back into line. In Cannon's view, given the truly fundamental importance of homeostasis in physiology, this was embarrassing. Today, physiologists no longer need feel embarrassed. But some have certainly been astonished by discoveries made only this century.

We now know that regions of the brainstem, which lies just above the spinal cord, contain cells that drive all kinds of automatic bodily processes, including heart contraction, breathing, digestion, the dilation of blood vessels, sweating, swallowing and vomiting. The activity of these cells is also directed by signals coming from a brain region lying close by – the hypothalamus, which is dedicated to homeostasis.

How do they get the information they need to do their job? Via a system of senses that together come under an umbrella term coined by Sir Charles Scott Sherrington, no less: 'interoception' – inner-sensing.[4] By enabling homeostasis, these inner senses allow us not simply to stay alive, but to push our bodies right to their limits, and survive.

It's 6 June 2012; the waters off the island of Santorini, Greece, are like crystal. The Austrian freediver Herbert Nitsch is preparing to mount an attempt on his own world record for the deepest dive on a single breath. Under the official No Limit rules, Nitsch will strap himself into a sled that will carry him down, and then back up to the surface. His existing record, which he set five years previously, stands at 214 metres. His goal today is 244 metres – as deep as a seventy-storey skyscraper is tall.

To prepare for a dive like this involves intense training. A healthy man can fit about six or seven litres of air in his lungs. By means of extensive exercises, Nitsch has increased his own regular lung volume to ten litres, which he can extend even further, to fifteen, using a technique called buccal pumping.[5] When he does this, it looks as though he's swallowing the air, taking down gulp after gulp.

He knows that to complete this dive, he'll have to hold his breath for about four and a half minutes. But as the record-holder at this, with a time of nine minutes and four seconds, this should not be a problem.

Still, Nitsch knows that as he descends ever deeper, he'll have to minimise his body's oxygen demands, and that will mean staying perfectly calm. What Nitsch terms 'body awareness' – feeling in tune with the state of his body, and also being in strong mental control of it – will be essential, because the dive will profoundly challenge his brain's ability to keep his internal bodily environment within the fairly narrow ranges that are essential for health, and for life itself.

The principal conscious homeostatic signal that he knew he'd have to overcome was the urge to breathe. You could try it now. Hold your breath, and focus on what it feels like . . . When you get to the point where you just can't hold it any more, the cause is not a lack of oxygen but rather the sensing of a rise in carbon dioxide levels in your blood.[6]

Carbon dioxide builds up as a waste product of energy-generation in our cells. We have to get rid of it, because it reacts with water in the blood to increase the concentration of hydrogen ions. The 'H' in the pH scale, which is used to measure how acid or alkali a solution is – and which, incidentally, was created by a Danish chemist working at the Carlsberg Laboratory, which was set up largely to serve the Carlsberg brewery – is for 'Hydrogen'. More hydrogen ions equal greater acidity, which alters biochemical reactions in brewing, potentially ruining a lager, and in blood, potentially killing you.

Chemical receptors that respond to hydrogen ions keep tabs on your blood. On the surface of the medulla, in the brainstem, are also receptors that monitor carbon dioxide in cerebrospinal fluid, the liquid that bathes the brain and spinal cord. When signals indicate that carbon dioxide levels are rising, you are prompted to inhale. If you're unable to (perhaps because you're attempting a world record-breaking freedive) you'll be sent ever sterner warnings to comply.[7]

But again, says Nitsch, you can train yourself to become less sensitive to these signals. His principal training regimen for this takes place not in water, but on his sofa. 'I developed this to do before any serious dive – whether it's a competition, training or even a vacation. So I lay on the couch, watching a lazy programme that's easy to follow, like *The Big Bang Theory*, so I have some distraction, but I don't have to really focus on it. Then I do breath holds. So I exhale, and hold my breath on empty lungs. I might start with just one and a half minutes. Then I breathe normally, then do a breath hold again, slowly but surely increasing the time. I do this for an hour at a time.'

By training his CO_2 tolerance in this way, he can extend the time before he feels the urge to breathe.

It's not only a rise in CO_2 that makes you want to gasp, however. To hold your breath, you have to keep your diaphragm muscle still, and mechanoreceptive signals from a stationary diaphragm contribute to the feeling of ever greater discomfort. With practice, though, you can learn to overcome this discomfort, too (Nitsch assures me).

Generally, access to oxygen is not a problem for us, and we're prompted to breathe well before we come close to running out of it. But we do still monitor oxygen in our blood. It's long been known that this responsibility lies with the carotid bodies – clusters of sensory cells in the carotid arteries, which supply the head and neck with blood. However, only in 2015, around ninety years after Walter Cannon wrote about how carotid bodies must 'taste the blood',[8] did Nanduri Prabhakar and his colleagues at the University of Chicago identify the long-sought oxygen sensor. They discovered that the oxygen-sensing system depends upon an enzyme called heme oxygenase-2. When oxygen levels in blood fall, this enzyme's activity stops. Sensory neurons in the carotid bodies register this, and send signals on to the medulla, causing a rise in the breathing rate.[9]

For Nitsch, preparing to dive, or for you (well, me) sitting here at my desk, other inner senses are also vital. Baroreceptors – mechanoreceptive stretch sensors – send sensory signals every time your heart contracts. They keep your brain informed about the frequency and also the strength of your heart beat.[10] This is critical, both for general blood-pressure control and also for the 'baroreceptor reflex'. If you're lying down to read this book, and you were suddenly to jump up, your blood pressure would drop very rapidly, but (it is hoped) also very temporarily, because baroreceptor signals would tell your heart to pump harder, to get more blood up to your brain.

Elsewhere in your circulatory system – in some large veins, and in parts of your heart – another set of baroreceptors monitor general blood pressure. Their signals help your brain to regulate the amount of water in your blood, and so its volume – and so the pressure that it exerts on the vessel walls.

In 2018, Ardem Patapoutian and his team solved a 100-year-old mystery, reporting in the journal *Science* that they had discovered the ion channel proteins involved in baroreception. They turned out to be Piezo2 (the very same protein that is important in touch and proprioception) and also a related protein, called Piezo1. That same year, the lab reported that another protein, called GPR68, functions as a receptor that senses blood flow.[11]

When Herbert Nitsch trained his lungs to accommodate as much air as possible, work in Patapoutian's lab suggests that it was mechanoreceptors expressing Piezo2 that also sensed his lung expansion, and told him when he had to stop. Not that Nitsch always listened. 'At first, it feels uncomfortable to hold your breath on extremely full lungs, because you feel the pressure, and it's not nice,' he says. 'But if you keep doing it, you can come to be comfortable with more air in the lungs. This is something you can train.'

For Herbert Nitsch, seeking that day to extend his world-record dive depth by thirty metres, all of these sensory systems were tested. Twenty-one minutes before the scheduled dive time, he slipped into the water, in his wetsuit, his nose clip and goggles at the ready. The crew of his support boat, plus assembled photographers and reporters, were looking on. Safety divers, equipped with SCUBA gear, who would go to his aid should he need it, readied themselves.

At forty-two, Nitsch had been building up to this dive for more than a decade. It all started when, aged twenty-nine, he booked a ten-day diving holiday to Egypt, and the airline lost his diving gear. Forced to snorkel, he found he could go ever deeper, for longer, doing much of what the divers could do with their breathing apparatus. When a friend asked how deep he could go, he managed thirty-four metres. A couple of months later, this same friend called him up to say he'd been only two metres off the Austrian freediving record. 'And he explained to me that really what I had been doing was not snorkelling, but freediving,' Nitsch recalls – and his passion was born.[12]

To June 2012. At this point, Nitsch holds thirty-three freediving records. He's prepared to withstand pressures that will shrink his lungs to the size of lemons, and to drop at seven metres per second to the blackest depths, with the knowledge that he *cannot* breathe. The physical preparations have been one thing. But to make it, he knows he has to keep his body's oxygen demands down to the minimum – and to be supremely relaxed. His technique for doing this came naturally to him, he says:

> Imagine the situation – first, you plan for this event for years, and then you have cameras, spectators, and everybody is concentrating on you, and you have to be like, 'I don't care.' Like you're waking up on a lazy Sunday and just turning back around. Even though you are in the middle of all that, you have to be that way. So what I do is I more or less depart my body, and see the whole thing from a bird's eye view. This is a lot easier when you have your eyes shut. Then you imagine the whole event, and you are in the middle of this frame, but because it's not you, you are just a spectator, you are not excited. You are an observer and therefore, you are relaxed.

Nitsch was strapped into the long sled. It was attached to a rope, and weighted, to take him down. In the base were two SCUBA tanks, whose valves would open automatically when the rope ran out, to release air into a section above his head to lift him back up.

Twenty seconds before dive time, his support divers surround him. He finishes packing air into his lungs, raises a gloved hand to indicate that he's ready – and he starts to sink.

One hundred metres . . . 115 . . . 120. A camera strapped to his body shows his face, set; his wetsuit, rippling. Up on the support boat, someone shouts out, 'Three minutes!'

The rope runs and runs, all the way down to 253.2 metres. Then, suddenly, the SCUBA tanks are opened, and he's on the ascent. At twenty-four metres, the sled will automatically stop, for a one-minute decompression break. This will be essential to eliminate nitrogen, and other gases, that have dissolved under pressure in his blood. If it doesn't happen, the gases can form bubbles, which if they get to the brain can block the blood flow.

The cameras show that when he's about 100 metres from the surface, he is no longer awake. The safety divers waiting at the twenty-four-metre point immediately assume that he's blacked out from a lack of oxygen, and rush to get him to the surface. Nitsch, alert now, tries to pull on the rope to stop them, but it's too late.

At the surface, he grabs an oxygen mask and goes back under, to nine metres, hoping this delayed decompression stop might be sufficient. But about fifteen minutes after the dive began, he's in real trouble. 'I felt a lack of feeling of control in my muscles,' he remembers. Immediately, he was rushed to shore in a speedboat. By the time he got to hospital, he'd suffered a series of mini-strokes.

A build-up of nitrogen in the blood can cause 'nitrogen narcosis'. It's a similar feeling to being drunk on alcohol, Nitsch explains. That, coupled with insufficient sleep in the build-up to the dive, meant that he fell asleep under the water . . . Even asleep, he didn't inhale. But the decompression sickness that resulted from his missing the planned stop was disastrous. His memory and his movement were both affected. He was told that he would be restricted to a wheelchair, in need of care.

Just seven months later, however, he had recovered enough that he was medically cleared to dive once again. Now, eight years on, he says he still has some difficulties with balance and coordination and also articulation. But under water, he feels just as he did before. He regularly freedives, but for pleasure rather than to compete.

Few of us might want to go freediving – at least at anything close to Nitsch's level. But his training procedures show us that the functioning of these inner senses can be modified. And more than that, though their signals are used by the brain to automatically control the body, we can become intuitively aware of some them, such as our heart beat.

The implications of this are only now being recognised. Later in the book, I'll look at the startling new research linking these processes to how we feel emotion. But the implications for our physical and mental wellbeing can be critical, too.

★

One of the 'unconscious' signals that Herbert Nitsch had to be aware of was his own heart beat. If it became too fast, he'd use too much oxygen, too quickly. He did at least know, though, that when he held his breath, his heart beat would slow down automatically. In fact, when he does this during his sofa-CO_2-training, his heart beat will often drop below the level at which a pulse monitor can detect it. 'Below about ten beats per minute, I get an error message!' he tells me. Many of us do, of course, use pulse monitors in smart watches or fitness trackers to monitor our heart rate. (Personally, I'm always horrified at the lingering effects of alcohol . . .) But what if I asked you to count your heart beats without feeling for a pulse, or looking at your smart watch? Could you do it?

It turns out that about 10 per cent of us can do this really well, 5 to 10 per cent are just terrible, and the rest fall in between. These figures come from studies in which people are asked to count their own heart beats over varying intervals of time while a pulse oximeter takes an accurate measure. The researcher then compares the two sets of results. Another test of 'cardiac interoception' involves playing a person a series of beeps and asking them whether those beeps were in or out of synch with their heart beat. About one in ten people perform really well at this, though 80 per cent just can't do it.[13]

At the University of Sussex, Lisa Quadt is working with colleagues on assessing heart-beat-sensing, and researching the potential implications of being good or bad at it. When I visit the labs, and she asks me if I want to have my ability tested, I can't resist taking her up. After sitting down, I slip the index finger of my non-dominant, left hand into a plastic pulse oximeter, which is wired up to her laptop. This device will accurately measure my heart rate. 'We'll do six runs,' Lisa says. 'The computer will say "start", then you count your heart beats until it says "stop". Then say out loud what you think.'

I follow her instructions, repeatedly counting and reporting back. It turns out that this is something of a special skill for me. My accuracy is 0.97. Almost perfect. At first glance, this might seem like a trivial fact. Whether you can perceive your heart beating or not, it is beating, and your brainstem is in control, and that's all that matters – right?

Well, it turns out, it matters for how we exercise. And, as I read the research on this, my new-found inner-sensing smugness fades.

Exercise is good for you – of course. So, if you're fit enough, going for a five-kilometre run at ten kilometres per hour, say, is better than going for a two-kilometre run (or jog) at seven kilo-metres per hour. We all need to build up to physical fitness, however. And as we all know, this is easier for some people than for others. I always thought this variation had to be to do with base-level fitness: someone who walks a bit more than another, say, would start from a slightly better place. No doubt, this is true. But a German study published back in 2007 revealed that inner sensitivity matters, too.

The team gave thirty-four participants a test similar to the one I did for Lisa Quadt and then asked them to jump on an exercise bike for fifteen minutes. They were told to pedal as they liked. The team found that the people who were *good* at perceiving their heart beats showed significantly smaller increases in heart rate and blood flow and covered a significantly shorter distance.[14]

The two groups were matched for physical fitness, so that wasn't the reason. The only conclusion the researchers could draw was that those who can accurately sense the rate of their own heart beat are more sensitive to physical load: they *feel* the physiological pressures more, so they don't push themselves as hard. In contrast, the rela-tively numb bad-perceivers cycled faster, because they felt it less. People who are good at heart-rate-sensing could therefore find it harder to get fit and build their fitness, because exercise is more off-putting.

The relationship between inner-sensing ability and fitness isn't quite that simple, however. Research is also showing that more heart-beat-sensitive adults and kids tend to be fitter. And there are a few potential interacting reasons for this.

One is that sensitive types might not choose to cycle hard in a lab experiment, but in the real world, they are better able to push themselves within their limits than less sensitive types. In other words, they could have better self-control when it comes to exercise. That could easily make for fewer injuries, less significant pain, and a more successful, achievable exercise regimen. Also, people who do regular exercise could be improving their heart-rate sensitivity. Certainly, overweight and obese people tend to be poorer at this.[15]

Still, weight and fitness levels aside, exactly what makes some people more heart-beat sensitive than others isn't clear. We know that there's a spectrum of mechanoreceptor sensitivity, and since baroreception depends on mechanoreceptors, this could help to explain it. But the state of the nervous-system pathways carrying information between the heart and the brain will play a role, too.

In fact, if these pathways are in good shape, you can expect all kinds of benefits – everything from better cardiovascular health and lower levels of inflammation to improved working memory and less stress.

One of the most important nerves in your body is the vagus (wandering) nerve. This nerve carries sensory information from the heart (and lungs and digestive tract) to the brain.[16] It also carries 'calm down', 'rest and digest' signals back down from the brain to all these regions. The stronger the activity of your vagus nerve, the more readily you can enter this state once a threat has passed. Strong activity is known as good 'vagal tone'.[17]

You can check your own vagal tone quite easily. Simply count your pulse (you can do this directly, with a finger on your inner wrist). Does it speed up a bit when you're breathing in, and slow down a little when you're breathing out?

If so, that's good news. This test reflects the vagus nerve's functioning, in allowing more freshly-oxygenated blood to circulate when you're inhaling, but also in slowing the heart's tendency to race when you exhale. The bigger the difference between the two, the higher your vagal tone.[18]

Vagal tone is at its peak in childhood. By adulthood, it's about as variable as height. It's thought that genes account for about 65 per cent of this variance. But lifestyle influences it, too. People who are overweight and do little exercise tend to have low vagal tone, while exercise helps you to train it.[19] It's even been claimed that this is a key reason why exercise is so good for you. And, as better vagal tone means your body can calm down rapidly after a threat has passed, it benefits both your body and your mind.

People with a strong vagal tone are better at regulating their blood glucose levels (which can reduce the risk of developing diabetes, or help those with type 2 diabetes to keep their levels in check). They are also less likely to suffer from a stroke or cardiovascular disease.[20] This could be partly because one of the vagus nerve's jobs is to dial down inflammation. Inflammation is a critical part of the immune system's response to an infection or injury. But if it persists at a background level over long periods of time, it can damage organs and blood vessels.

American neurosurgeon Kevin Tracey has pioneered the electrical stimulation of the vagus nerve to reduce inflammation. He's found that this therapy – delivered via an implanted pacemaker-type device – can effectively treat rheumatoid arthritis, an autoimmune condition.[21]

Tracey expects that his patients will need their stimulation therapy for life. But for the rest of us, if we can find ways to strengthen our vagal tone, the promise is not only that our brains will get better-quality sensory data in from organs including our heart, but that our 'calm down' process will work better, too, and we'll be in a rest and digest state for longer. Certainly, there's evidence that people with better vagal tone also report less anger and are better at controlling their emotions.[22] They may also benefit from a better working memory – the type needed to hold and manipulate information in your mind.[23]

There's some evidence that regular meditation might help, by encouraging your brain and body actively to engage the calm-down, 'relaxation' response. Marcelo Campos, a lecturer at Harvard Medical School, is just one physician who notes that healthy sleep, meditation and also mindfulness can all influence your heart-rate variability. But physical exercise seems to be an especially powerful way of improving it. Even in patients with chronic heart failure, high-intensity interval exercise has been found to increase vagal tone, and to reduce unhealthy irregularities in heart beat.[24]

Thanks to Aristotle, we often think of our senses as providing information only about what's around us. What the chapters in this second part of the book have made clear already, I hope, is that our own internal sensory world is intense and complex – but not inaccessible.

We can, and do, tune in to sensory signals about what's happening inside our own bodies. And we depend upon these senses for everything from Nitsch-style achievements to feats that are far less extreme, but no less important for us, as individuals – to complete the 'Couch to 5k' challenge, say, with all of the flow-through bene-fits for inner-sensing that physical fitness brings.

As we also know now, too, even some of the Aristotelean senses don't fit squarely into the 'external' category. Smell and taste recep-tors are blind to the internal/external border, doing their job of detecting chemicals of interest, wherever they are. And in the next chapter, I'll look at another type of sensing that disregards that distinction, too.

9

Temperature

Why cats and dogs make us happy

Google 'the hottest', and a pretty mixed bag of popular searches will pop up: the hottest place on earth, the hottest planet in the solar system, the hottest dog, the hottest chilli peppers, the hottest celebrities in the world.

The hottest *dog*?

Maybe you already know that it's a YouTube channel . . . This was news to me. It, at least, has nothing to do with temperature-sensing. But everything else on that list certainly does.

Remember, though, how, a century ago, physiologists were embarrassed at their lack of understanding of homeostasis? For sensory scientists, this is the modern equivalent. As one recent review put it: 'Some physiological sensations have clear origins and unfold in predictable ways, but thermosensation is not one of them.' However, we do at least now understand why a bird's-eye chilli, a steaming bath, a summer's day in Delhi and – yes – even some celebrities are 'hot', and why an assortment of other things, including peppermints, say, and England in late autumn, are decidedly cool.

Even more remarkable, though, is new research revealing links between temperature-sensing and how we think and feel. This sensory system, it turns out, is not only practically useful. It also has a raft of impacts on our psychological wellbeing, and, potentially, our physical health.

In this chapter, then, we'll look at how this class of sensing happens and what it means for us as individuals (because we don't all sense temperature in the same way). We'll also consider the often surprising ways in which it affects us, whether or not we're conscious of its influences – and how we might manipulate the system, for our own good.

*

To understand how we sense temperature, it's helpful to go back in evolutionary time – right back, in fact, to the common ancestor of both a type of nano-plankton and the entire animal kingdom.

What this ancient ancestor was really like isn't clear. But it evolved sensory adaptations that have been passed down to us. How do we know? Because if you were to delve into your DNA, you'd find genes for a class of sensors called transient receptor protein channels (TRP). Delve into the DNA of a modern choanoflagellate nano-plankton floating deep in Antarctic waters, and you'd find some incredibly similar genes.[1] These are truly ancient sensors, and though they have evolved to have a dizzying array of uses throughout the animal kingdom, in us a sub-family known collectively as 'thermo-TRPSs' allows us to sense temperature. These receptors are found in all kinds of organs in your body, including your brain, liver and digestive tract, as well as, of course, your skin.

Individual thermo-TRP sensors, which are found in the membranes of free nerve endings, react to different temperature ranges. *Exactly* what they respond to, and what we perceive as a result, is still being explored. But the latest thinking holds that they function like this:

Warmth/heat-activated sensors:[2]

- TRPV1 (heat/dangerous heat). These sensors respond to temperatures above about 43 °C, acids, capsaicin (the 'burning' ingredient of chilli peppers) and tarantula venom. This was the first temperature channel to be discovered in people, in 1999. Its activation can cause pain.
- TRPV2 (dangerous heat?). The second member of this family to be identified, it can be activated by extreme heat, above about 52 °C. However, whether we use it predominantly for sensing high heat or something else entirely is debated, partly because it's been found to have other roles within the body (in relaxing the stomach and intestines, for example).
- TRPV3 (warm). These receptors are activated by temperatures in the range of about 32 °C to 39 °C, and also by vanillin, camphor, cinnamaldehyde (found in cinnamon sticks) and frankincense.
- TRPV4 (warm/neutral). It responds to temperatures in the range of about 27 °C to 35 °C.

Cold-activated sensors:[3]

- TRPM8 (cool/cold). Our chief cold sensor, it responds to temperatures below 26 °C and to menthol (found in peppermints and some mouthwashes) and eucalyptol (an oil from eucalyptus trees, also used in some mouthwashes and skin creams).★

- TRPA1 (cold, and dangerous, painful cold). These receptors start to react when the temperature drops below about 17 °C. TRPA1 is also activated by the chemical in onions that makes you cry and by the compounds that give wasabi its punch. These chemicals don't seem cold, however . . . So is TRPA1 really a painful cold sensor? There's still some debate. However, since mice bred to lack the TRPM8 receptor can start to sense a cold surface when its temperature drops below about 10 °C, mammals clearly do have another cold sensor.

When you know what these individual types of receptor respond to, the reasons why a curry, a heatwave, a drop of strong acid and a cup of coffee are all 'hot', and a peppermint and a glass of water are both 'cool', become completely clear. Still, perceiving a cup of fresh coffee or a summer's day in Delhi as being hot makes sense. Why should a mouthful of a chilli-laced salad stimulate the same sensory receptor?

It's thought to be because the chilli pepper has evolved to hack into the mammalian heat-sensing set-up, for its own ends. Capsaicin can be viewed, then, as a chemical weapon in the fight against being eaten. (Though when it comes to humans, it's not a particularly effective weapon. There's even evidence that we've domesticated the chilli plant on at least five independent occasions; *that's* how much humans like it.[4] Still, we are an exception. The only other mammal known to seek out the plant is a Chinese tree shrew, and it has a TPV1 mutation that makes it less sensitive to capsaicin.[5]) Other species, too, target this very same receptor. David Julius, who has led a vast amount of work on temperature-sensing at the University

★ TRPM8 is also the cold sensor for cold-blooded animals, such as frogs – but in these species, it responds to an appropriately lower trigger temperature. (Frogs are sensitive to menthol too, however).

of California San Francisco, has found that certain spider toxins bind with it, too.[6]

If you do consume something at room temperature, and feel a sensation of heat or cooling in your mouth, throat or lips, that's because chemical stimulation of temperature receptors has caused signals to race along the trigeminal nerve, which has branches in the eyes, nose, tongue and mouth, to your brain. 'Trigeminal sensing' is not smell, and it's not taste.[7] It allows you to detect acrid and also caustic chemicals, such as ammonia in bleach, as well as chemical heat, as from chillis, and cool, as from a peppermint perhaps.

Other parts of your body are also sensitive to the heat of chilli or the cool of something like menthol, of course. Rub a menthol ointment on your forearm, and you'll feel a cool tingling in that spot. But your lips are especially sensitive, and that's because these various receptors aren't present in the same density throughout your skin. That's something that's very easy to explore for yourself. Right now, as I'm typing, it's January in Yorkshire (it seems an awfully long time since I had to switch on that fan in Chapter 1). My house is heated, but still . . . If I lay the fingers of my right hand on the back of my left hand, my fingers feel a little chill. If I now press my fingers against my lips, the cold jumps out.[8]

What complicates this response a little is that the temperature of the body part that's doing the temperature-sensing matters, too. If you were to put one hand in a bucket of cold water and the other in a bucket of hot water, then put both hands in a bucket of luke-warm water, because of the pattern of previous receptor stimulation, your right hand would tell you it was hot, whereas your left hand would argue for cool. Soon, though, the system would settle down, and you'd accurately perceive lukewarm.

Though your fingertips and palms aren't as temperature-sensitive as your lips, they're still competent. And the kinds of temperature information that they can garner can provide you with all kinds of insights. For example, if you were to close your eyes, outstretch your palms, and have someone place a block of wood on one and a block of metal on the other, you'd know which was which just by the fact that metal lowers your skin temperature more rapidly than wood, and so feels colder.

This kind of discrimination isn't the most important job for our temperature sensors, however. They have two principal roles. The first is to provide the information necessary for your ability to maintain a core body temperature of about 36.5 to 37.5 °C. Since your body and brain stop functioning properly outside this range, this is a key-worker task, if ever there was one. The second is to scream the alarm if something dangerously hot or cold is touching you.

The hypothalamus, your master-controller of homeostasis, itself contains temperature receptors.[9] But signals from those present in the skin are also important for regulating your body temperature. If you're sitting at your desk, or on the sofa, and suddenly perceive the room to be too cold, you can go and grab yourself a jumper. If you're feeling uncomfortably warm, you can peel off a layer. Yes, your brain has various temperature-regulating options at its disposal, but if your muscles can easily help with this, you're going to feel the impulse to obey its unconscious will.

If the temperature in your office or your living room does alter, even a little, you'll register it. However, this wonderful sensitivity isn't coupled with an accurate insight into the degree of that change – and we tend to over-estimate it. If you open a window just a crack, and someone shouts out that it's suddenly freezing, this is why.[10] This exaggerated response is of interest to sensory scientists, of course, but also to another, perhaps less predictable group.

A sudden 'chill' – a perceived 'plunge' in temperature – is often taken by believers to be a sign that a spirit has entered the room. (A *draughty* room, no doubt. Ghosts are of course known to favour crumbling castles over temperature-controlled eco-homes with snug-fitting windows and doors.)

Parapsychologist Ciarán O'Keeffe at Buckingham New University is interested in this effect. In their book, *In Search of the Supernatural*, O'Keeffe and Yvette Fielding describe an investigation at Hever Castle, in Kent. Once the home of Anne Boleyn, it is 'a castle straight out of a fairy tale'. O'Keeffe prepares to position some temperature data loggers around the grand dining-room. Reports of ghost encounters at Hever Castle have included mention of cold spots, he notes:

However, it's worth bearing in mind that the correlation between hauntings and temperature is made much of by the media and possibly already in the forefront of someone's mind . . . Here in Hever Castle, I'm particularly conscious of the structure of the building and how there are some very obvious channels for air, like the fireplace . . .

Add a cool draught from the fireplace to a hefty dollop of belief and a 'ghost' encounter could easily be cooked up.

It's not just slight body cooling that we react to, however. The mercury doesn't have to rise by much to send you careering off in the other direction.

Tom Hardy, Idris Elba and Gerard Butler all feature on the 'definitive list of the world's hottest males' of 2020 – according to the radio station Heart, at least.[11] Fitness coach Joe Wicks also made the list. (And frankly, as one of countless parents who lunged and air-punched their way through Wicks' YouTube PE lessons every school morning during the coronavirus lock-down, for me there *was* a draw, beyond staying fit.)

We commonly use the term 'hot' to describe someone who inspires even a very mild state of sexual arousal. But of course, *they're* not hot. It's us who's getting a little warmer, in parts.

To explore this, we need look no further than a study involving semi-naked people and thermal-imaging cameras. Male and female volunteers sat (individually), naked from the waist down, to watch everything from *The Best of Mr Bean* to Canadian tourism travelogues to pornography. All the while, one of these cameras was aimed at their genitals.

Discreetly, from another room, the McGill University researchers monitored the participants' body-temperature changes, down to within one-hundredth of a degree. The results showed a spike in genital temperatures, caused by increased blood flow, only when they were watching porn. On average, for both men and women, these temperatures rose by about 2 °C.[12]

If that spike had happened body-wide, they'd have cracked a fever. So, yes, within a certain region, they were feeling very hot indeed.

*

Over-reacting to even small changes makes sense. When it comes to managing our body temperature, there's very little margin for error. We need to be right on any change – whether it's caused by a process outside or inside the body – as it happens. But the way that our temperature-sensing system works means that there's scope for some other, much stranger effects.

Temperature receptors are indeed found on the endings of nerve fibres whose main job seems to be to transmit this information to the brain. However, that's not the whole story, as a brilliant experiment dating back to the mid-nineteenth century demonstrates.

This is one that's pretty easy to try for yourself. Or, ideally, with a friend.

1. Find two one-pound coins (or any two identical coins, in fact).
2. Put one in the fridge for ten minutes or so.
3. Meanwhile, warm the other one in your hand to skin temperature.
4. Sit in a chair, tip your head back and put the warm coin on your forehead. Concentrate on how heavy it feels.
5. Now swap it for the cold one. Does it feel heavier?

It was a German physiologist and experimental psychologist called Ernst Heinrich Weber who reported, in 1846, that when a cold coin was placed on the forehead of a volunteer, they reported that it felt as heavy as, if not heavier than, *two* warm but otherwise identical coins.

Weber's wonderfully puzzling discovery somehow languished in obscurity for over a century. Then in 1978, two American researchers, Joseph Stevens and Barry Green, decided to test it for themselves. They discovered not only that Weber was right, but that it wasn't just the forehead that could be tricked in this way. When cold and warm weights were placed on volunteers' forearms, the same kind of effect happened. And it was dramatic. A cool ten-gram weight felt as heavy as a 100-gram weight at skin temperature.[13]

How could this be? Further research has revealed that the endings of some C fibres that mostly send on touch signals also respond weakly to cold. Because their main role (and one that they're very good at) is signalling touch, when the brain receives signals from them, it thinks these signals are indicating a degree of pressure, rather than cold, and the cool coin is perceived as being heavier.

In fact, Stevens and Green have made all kinds of fascinating temperature-related discoveries, including:

- Cooling decreases the umami taste but increases the salty taste of monosodium glutamate.[14]
- The same stimulus feels less rough when skin temperature drops below normal (32 °C).[15]
- Cooling the skin dulls spatial touch acuity (your ability to detect separate points of contact), but if you're touching something that itself is cold, this ability is enhanced.[16]

The coin illusion isn't like the chequerboard illusion from Chapter 1, however. It is not an example of our brain fibbing to us for our own good. Rather, it's an artefact of the way these sensory data get to the brain. It helps to show, though, that some apparently bizarre sensory sensations (a cold coin feels heavier – *really?*) can be both genuine and totally explicable.

And if we're talking bizarre and only recently explicable, it's worth taking a closer look at TRPV1, our high heat sensor. Because in some animals, it's been adapted to do something quite extraordinary.

The common vampire bat has a squished nose, Spock-like ears and also razor-sharp fangs, which it uses to 'practise haematophagy', as a zoologist might put it – or to 'feed on blood', as anybody else would. This bat, which lives in the Americas, likes to creep up on sleeping animals. Calves are a particular favourite.

How does it locate its prey in darkness? Smells and sounds help (it even has brain cells that respond specifically to breathing sounds[17]). But this bat also comes equipped with a sophisticated body-heat-sensing system. Around its 'nose leaf', as it's called, are pit organs, which contain infrared sensors. These sensors can detect the heat radiated by warm-blooded creatures. But they can do even more than that. This system is so gloriously, macabrely sensitive to infrared that it can even tell the bat where, on an unwitting animal, a blood vessel is lying particularly close to the surface – and so is the perfect spot to bite.

In 2011, a team led by David Julius reported that the receptor essential for this IR-sensing is none other than a modified form of TRPV1. In the rest of the bat's body, as in ours (as in all mammals, in fact), the standard form works as a high heat detector. But in those pit organs, a shorter version responds to the slightly cooler 30 °C temperature of blood pulsing away behind skin.[18]

Though other animals can sense the body heat of prey, so far the vampire bat is the only one known to use TRPV1 to do it. We just don't have this form. However, our own TRPV1 receptors can be tweaked to respond to this kind of temperature. If you've ever stepped into the shower with sunburn, you'll know that your usual shower temperature will suddenly feel burning. Why? Because chemicals released by your immune system in response to sun damage lower the threshold temperature at which TRPV1 receptors will start firing, bringing it down from about 43 °C to 29.5 °C. This helps to stop you doing more harm to your severely damaged, vulnerable skin.[19]

However, some of us are generally more, or less, sensitive to hot temperatures, capsaicin or both. At least six variations in the TRPV1 gene have been found in people, which help to explain this. Yes, just as there are individual differences in all the other senses that I've looked at so far, there are fundamental variations between us in this system, as well.

One genetic variation in particular causes a markedly stronger response to capsaicin. This could explain why some people just can't stand chilli-hot food – or at least react more, flushing and sweating more than their dining companions.[20]

Right at the other end of the temperature-experience spectrum are the people who take part in chilli-eating contests. A jalapeño chilli – the sort you might get on a pizza – scores between about 2,500 and 8,000 on the Scoville Heat Unit scale, the standard measure of chilli heat. (This is a measure of how many times the capsaicinoids extracted from a precise weight of dried pepper have to be diluted with their own weight in sugar water to stop them from seeming hot, as judged by a panel of tasters.) A jalapeño can make your mouth sting. But it's a mere firecracker compared with the atomic bomb of the Carolina Reaper, which in independent lab tests has topped 1.6 million Scoville units.[21]

Some people willingly eat Carolina Reapers, the world's hottest chillis, in contests, either on their own or in a coating for chicken wings. Because eating just a bit of one simulates intense heat, physiological cool-down orders are immediately despatched from the hypothalamus.

Though few of us are brave (crazy?) enough to munch on Carolina Reapers, we all know what this response roughly entails. Lie on a lounger, under a blazing sun, and you'll start to sweat – which, when it evaporates, will cool you down. Your heart rate will also rise, to push more blood towards your skin; at the same time, blood vessels here will relax, allowing up to a doubling of blood flow, so enabling more heat to radiate out of the body, into the air – and giving you a red face in the process.

Moving around too much would only burn energy, releasing more heat, so (assuming you've removed whatever clothing is possible/appropriate) your skeletal muscles are put on the go-slow. You feel fatigue, a state that discourages you from doing anything but staying still. So you just go on lying there, 'relaxing' – again, almost against your will. (In fact, 'acts of free will are hostage to a host of inner body states,' notes Olaf Blanke, the Swiss-based neurologist mentioned in Chapter 7.)[22]

A little time on a sun lounger is one thing. But a Carolina Reaper signals an inferno-like heat. In response, the hypothalamus mounts the biggest cool-down offensive that it can. People who have eaten these chillis describe not just a prolonged, searing, gut-wrenching pain but also a horribly racing heart, and uncontrollable sweating. 'Vomiting felt like an exorcism,' one consumer has written, 'because my body was desperate to be rid of the demon inside.'[23]

Which begs the question: why do they do it? We'll look more closely at pain in the next chapter, but heat pain like this triggers a rush of adrenaline, and the body's own natural painkillers. For some people, the pleasure is worth the pain. Such 'sensation-seekers' crave stimulation, and a California Reaper certainly falls into that category. But perhaps you can see less extreme versions of this within your own family?

When my sons were little, the older one would always cover his ears during fireworks, while his brother, two years younger, would get as close as possible, shouting, 'More! Louder!' Five or so years on, guess which one rides his bike sedately, and which hurtles along mountain trails, at full pelt – and which requires five glasses of water to get through a chicken korma while the other douses his dinner with Nando's hot sauce at every opportunity? He feels the heat – his face goes red, he pants – but he *enjoys* it.

Extremes of temperature can do more than feel good, however. They can also improve our health.

On a January day in Oslo last year, I stood near the Opera House, which was covered in snow and ice, and watched people leaping from a raised platform into the fjord. They were surprisingly silent about it all. But as they jumped into that water, which had to be down at around 4 °C, their peripheral blood vessels would suddenly have constricted, the hairs on their skin would have stood on end in a bid to trap their body heat, their heart rate would have rocketed – and they surely started shivering.

Unsurprisingly, they wasted no time swimming to a ladder, clambering back up to the platform, and hurrying into a wooden sauna. The sudden heat would then have caused the warm-up brakes to screech to a halt, and before long, the entire system would have careered into reverse.

The Scandinavian sauna is famous, but humans have been using heat, if not both cold and heat, for thousands of years. The ancient Mayans of Central America used sweat houses for health as well as religious purposes. The Ancient Greeks, too, were fans of hot and cold baths.[24]

With the Romans, though, 'thermalism' became an obsession. Every town had a bathhouse. From the warm room (*tepidarium*), the bather would progress to the hot room (*caldarium* – quite like a sauna) and then the *frigidarium* (cold room). Some physicians were especially convinced of the health benefits of cold-water bathing, even in winter. The Roman scholar Pliny the Elder, who lived, according to Biblical accounts, around the same time as Jesus, wryly observes the extreme influence of one such advocate, a man known as Charmis of Massilia. The health craze that he helped to encourage meant 'we used to see old men, ex-consuls, frozen stiff in order to show off,' Pliny writes, adding that other Greek doctors persuaded healthy Romans to take 'boiling' baths.[25]

High heat or cold – and especially a switch between extremes of cold and heat – certainly challenges your cardiovascular system. Many doctors advise against it, because of the stresses of such abrupt changes on that system. When it comes to the type of sauna you might find in a gym, the medical advice is to use one only if you're fit enough – if you can walk for half an hour or climb three or four flights of stairs without stopping. If you can, there is some evidence that repeated sauna use lowers blood pressure and reduces the risk of fatal heart disease and stroke. This finding is from a study that followed, over twenty years, more than 2,000 men in Finland, a country where there are reportedly as many saunas as TVs.[26]

However, you don't need a sauna and a fjord to hand to enjoy the potential benefits of shifting between heat and cold. In sport, it's not unusual for coaches to use alternating hot/cold baths as a post-exercise general recovery treatment for athletes, though the results of scientific studies on this are mixed.[27] There are plenty of top athletes, though, who'd side with Charmis of Massilia. In 2017, when Andy Murray was preparing his painful hip for Wimbledon, he spent eight to ten minutes every day in an ice bath, and underwent another freezing immersion every evening. 'It might not be everyone's ideal preparation for a good night's sleep,' he told the BBC at the time, 'but fortunately I've got used to plunging myself into ice-cold water over the years and I don't mind it, I'm OK.'[28]

Switching between blood vessel constriction, caused by the sensing of cold, and blood vessel dilation, triggered by subsequently sensing warmth or heat, helps to stimulate blood flow and reduce swelling. There is more than anecdotal evidence that if you have an injury, this can help.

As we know, the Ancient Romans didn't only use heat and cold in their baths, however. There was also the *tepidarium*. Designed to be warm and pleasant, it certainly wouldn't have got the bathers' bodies into anything like the stunned state of the other two rooms. But that doesn't mean it wasn't having a potent effect.

On the face of it, sensing warmth may not seem that exciting. But that's only because we've under-appreciated it. Understand how it affects us, and how to work with it, and the promise is that you can improve everything from your work life to your mental health. Warmth is not the mild-mannered, unassuming cousin of heat and cold. In fact, it does phenomenal things to your brain.

When you're feeling pleasantly warm, it's true, you're in a comfortable zone. Your body doesn't have to be working to heat you up or cool you down. On this life-critical measure, at least, your hypothalamus can temporarily relax. No wonder, then, that being warm feels *good*.

However, what makes you, and somebody else, feel 'warm' can easily be different. And if there are rows in your house about whether the thermostat should be turned up or down, odds are there's a woman on one side of that argument and a man on the other.

In this case, though, the cause is likely to be less about genetic variations in temperature receptors and much more about basal metabolic rate. That's the amount of energy that has to be burned to keep the body functioning while you're doing nothing. Men have a metabolic rate that's up to a third higher than a woman's. This means they produce more heat. To get their skin temperature within the zone that both sexes consider comfortable (around 33 °C), they therefore need less heat input from their surroundings. So the room temperature that men report as feeling pleasantly warm in (about 22 °C) can be up to three degrees cooler than for women (who report preferring around 25 °C).

The guidelines for acceptable temperatures for office thermostats are based on a formula devised in the 1960s, calculated using – somewhat predictably – only the metabolic rates of men. Women will, then, tend to feel 'too cold' in an office that their male colleagues find perfectly acceptable. This isn't a minor problem. Work published by a team of German researchers in 2019 suggests that it makes a real difference not merely for comfort, but to the entire office's productivity.[29]

The team got male and female students to do maths and verbal logic problems in a room that was heated to anything between 16 °C and 32.8 °C. They found that when the room was warmer, the female students answered more of the problems, and got more right. A 1°C increase in the temperature was linked to an almost 2 per cent increase in correct maths answers for the females. The men, however, performed better in cooler room temperatures. But, crucially, in the warmth, the reduction in their performance was smaller than the enhancement of the women's. This suggests that to gain maximum productivity from a mixed-sex workplace, it's better to have an office set to be 'too warm' for the men, but right for the women. The same argument could be made for schools.

But while the actions of employers and educators can affect how warm we feel, with implications for how we function, we can manipulate our own warmth perceptions, too. One way to do this is with food. And I'm not talking about using an oven or a fridge.

The list of compounds that our TRPV3 warmth receptors responds to is enlightening, isn't it? There's cinnamon, a key ingredient of winter dishes, including mince pies and rice pudding, not to mention mulled wine. We even talk explicitly about cinnamon as being a 'warming' spice – and that, it seems, is exactly what it signals to our brains.

It's interesting to find vanillin, from vanilla pods, on that list, too. In the West, at least, we tend to like the scent of vanilla. It has even been claimed to have mood-boosting properties. There's been debate, though, about whether this might be because vanilla is a common ingredient in Western puddings, so we have learned to associate it with lovely treats, and feel good when we smell it for that reason – or whether something more biochemical might be going on. As vanillin triggers our warmth receptors, this does suggest that it's more than psychological.

Frankincense is worth an even closer look. A beautiful golden resin, tapped from the *Boswellia sacra* tree, it's known to anyone who's ever watched a Christian nativity play as being one of the gifts brought on the birth of Jesus by the three wise men. For thousands of years, it's been highly valued. (Its very name in English is a blend of the Old French *franc*, meaning noble, and incense.)

The Ancient Egyptians used frankincense while embalming their dead, but it's best known as an incense to be burned, and one that has been used in ceremonies important to many different religions. Pliny the Elder wrote in his *Natural History* about how in Arabia Felix (modern Yemen), priests took a tithe of frankincense as income, and also about methods that buyers should adopt to ensure that they weren't being duped:

> It is tested by . . . its stickiness, its fragility and its readiness to catch fire from a hot coal; and also it should not give to the pressure of the teeth, and should rather crumble into grains. Among us it is adulterated with drops of white resin, which closely resemble it, but the fraud can easily be detected by the means specified.[30]

He could easily be describing a drug . . . and in some other accounts, frankincense has been ascribed mind-altering properties. These reports intrigued Esther Fride, at Israel's Ariel University, who has researched the therapeutic use of cannabis, and who started to explore frankincense's biochemical effects. In studies on mice, Fride noted some anti-inflammatory properties, and it also seemed to boost the animals' mood.

Fride was well aware that many plants contain psychoactive ingredients. But she found that frankincense didn't trigger the kinds of receptors typically implicated in these effects. To her surprise, it was, though, triggering TRPV3. As Fride notes, the results imply that TRPV3 channels play a role in emotional regulation – and this mechanism may explain why frankincense has been so highly valued by so many religions, for so long.[31]

Why might warmth signals improve mood? As we already know, they mean that we're in a happy homeostatic state. But there could be more to it than this. The range of temperatures that we perceive as warm (32 to 39 °C) covers not just our own skin and bodily temperatures, but those of other people too. These receptors can indicate close contact with a parent, a lover or a child – a state of being that promotes our survival, or our genes' survival, at least.

I'll never forget, as a psychology student, being shown images of those now classic but also desperately sad 1950s studies of infant monkeys taken from their mothers and given the option of a cage-like fake 'mother' with milk, or a soft, cloth-covered 'mother' fitted with a warmth-emitting 100-watt light bulb (but without milk), to cling to. These infants spent as much time as they could with the warm, soft version. When others were given no choice – they were housed with only a cold, cage mother or a warm, cloth one – the second group grew up to have far fewer problems interacting with other monkeys.

Clearly, even the warm, soft mothers weren't much to speak of. They didn't respond to their infants, or actively soothe them. But just their warmth was something. For human babies too, from our earliest moments, we come to pair the essential presence of a carer with perceptions of bodily warmth. And there's a theory that, as we grow up, we maintain an unconscious link between physical warmth and feeling connected with other people. Our everyday language supports the idea. We certainly don't give beloved friends a 'cold' welcome, or a 'hot' one, but a 'warm' one. Neither do we talk about supportive, helpful individuals as being hot-hearted. They are 'warm-hearted' – not at all the kind of people to whom we'd give the 'cold shoulder'.

One really interesting feature of our warmth-sensing system is that while our other temperature receptors become sensitised to prolonged stimulation (which is why a steaming bath feels less hot once you've been in it a short time), warmth receptors don't. This means that you can keep on perceiving and enjoying physical human warmth as long as it lasts.

What if you're deprived of that warmth, and you're feeling lonely? Are there substitutes?

There's no end of work showing that pets are good for our well-being. Dogs and cats have a body temperature that's just a little higher than our own, and part of their psychological boost may be down to the physical warmth that they bring. But if you don't have a pet, there might be something else you could try.

In 2012, a team at Yale University asked a group of volunteers aged between eighteen and sixty-five to keep a diary of showers or baths and note down how they felt before and afterwards. The team found that lonelier people took more baths and showers, and stayed in them for longer. Their conclusion? That we unconsciously use warm experiences, such as baths, to dispel feelings of social 'cold' – of isolation.[32]

Since that study was conducted, groups have repeated it, and found similar results, while other groups have failed to replicate the findings, leading critics to query whether such a link really exists. However, work published in 2020 suggests that the reason for the discrepancies in results might be because these studies didn't take into account the ambient temperature – the warmth, or otherwise, of the day. On hot days, perhaps, lonely people may not feel such a need for enveloping water heat. On cold days, however, the US team found that people were more likely to crave social contact, and that this desire could be eliminated by warming them up. In this particular case, the effect was achieved with a heated back wrap (the sort advertised commercially for treating aches and pains).[33]

Other studies have found that people who report feeling warmer also say they feel more socially connected and agreeable. More work is needed to fully elucidate potential links between temperature-sensing and our social perceptions. But the work in this area to date is, at the very least, intriguing. Remember, also, from Chapter 5, that our caress receptors respond especially strongly to stroking at skin temperature. From our earliest hours, warmth, along with gentle touch, tells us that we're with someone else, that we're safe.

If warmth and even sauna-style heat can make us feel good, and cool isn't too awful, burning heat and freezing cold are a different story, of course. Now it's time to take a closer look at how our senses cause us pain.

10

Pain

Why heartbreak hurts

When Pavel Goldstein's wife was giving birth to their first child, she opted not to take any painkilling drugs. 'We had a really long delivery – around thirty-two hours,' he says, 'and she asked me to hold her hand . . . and talk less. So that's what I did, and it seemed to help.'

Anyone who's ever comforted a child with a scraped knee or even rubbed their own injured hand knows that physical contact can alleviate pain. But as a psychologist and neuroscientist at the University of Colorado, USA, Goldstein found himself driven to do more than simply note that holding his wife's hand helped her. It led him to embark on a series of studies to explore precisely how and in what circumstances touch works to reduce pain. This research, along with other investigations into how to mitigate pain – as well as studies of people who feel none at all, and yet others who experience a strange version of it – is opening up the long-mysterious world of some star players in the team of senses that we need in order to survive.

Just as we don't detect sounds, we don't detect pain. What we sense is damage, or the potential for imminent damage, to the body. This is 'nociception' – noxious stimulus detection. This sensory process can be separated out from pain perceptions, since one can happen without the other. However, when our damage detectors – our nociceptors – alert our brain to a clear threat to our bodily integrity, this usually results in pain sensations. Pain, then, is the feeling associated with the sensing of damage.

But how does it *feel*?

Partly because it is so private, pain is notoriously difficult to define and to measure. Aristotle could do no better than an 'unpleasant sensation'. Today, we still don't know precisely how the brain constructs the private experience of something hurting. (And, to be fair to Aristotle, a standard modern definition of pain is an 'unpleasant sensory and emotional experience associated with actual or potential tissue damage';[1] not a huge leap in clarification, then, in more than 2,000 years.) But in the twenty-first century, there has at least been what scientists in the field have termed a revolution in understanding the sensory biology of pain.

The French philosopher René Descartes is credited with being the first to identify pain (and temperature perception, for that matter) as an individual 'sense', in 1664.[2] In the early years of research, no one knew whether pain resulted from the activation of dedicated receptors and nerve fibres, or from the *over*-activation of fibres with an existing sensory role. Since, for example, 'blinding' lights and 'ear-splitting' sounds are painful, William James argued in the late nineteenth century in favour of the latter view. By 1903, however, Sir Charles Scott Sherrington felt able to declare: 'There is considerable evidence that the skin is provided with a set of nerve endings whose specific office it is to be amenable to stimuli that do the skin injury, stimuli that in continuing to act would injure it still further.'[3]

In 1906, Sherrington first wrote about 'nociceptors' – 'harmful stimulus detectors'. Since then, research in anatomy labs, on disembodied neurons, as well as in state-sanctioned minor torture chambers – in which willing volunteers have variously had their hands burned, their toes squeezed, their fingers pressed with ice and even balloons inflated inside their rectums – has confirmed the view that pain has its own dedicated sensory machinery.[4]

Damage sensors are found throughout our skin, internal organs, muscles, bones and joints, as well as in the membranes that surround the brain and spinal cord. They each respond to one, or more, of three types of damaging or potentially damaging stimuli: extreme temperatures (your TRPV1-expressing neurons are classed as nociceptors), hazardous chemicals and mechanical impacts – such as the slicing or crushing of cells.[5]

If you were to grab a pan from the oven with bare hands, signals speeding along A–delta-fibre damage sensors at about twenty metres per second would cause a sudden, sharp shock of pain – and you would drop the pan. After the initial shock, signals passing a little more sedately, at around two metres per second, along 'slow' C fibre pain neurons would result in a second wave of a more diffuse, burning or aching kind of pain, which wouldn't let you forget that your hand is injured and that you should protect and heal it. While some A–delta-fibre or C fibre nociceptor endings respond to high heat, say, or intense pressure, others are 'polymodal' – they can respond to multiple dangers. Precisely what will trigger them to fire off signals depends on which receptors are present in the membrane. Some receptors respond to extremes of heat or cold, while others signal the presence of dangerous chemicals, or physical damage.[6]

The chemicals that can cause pain do so for a few different reasons – because they happen to bind to a receptor with a different principal job (as is the case with capsaicin and TRPV1); because they can actually damage cells, or because they are signalling that cells have been damaged.

Acids can certainly damage cells, and we, along with other animals, have a suite of acid sensors.[7] There has to be quite a shift in pH for TRPV1 to detect an acid, and sound the alarm. But other detectors, including acid-sensing ion channels, can register even slight changes. Some nociceptive neurons have a suite not only of different types of damage sensors, but also of acid sensors.

Alkalis can also cause a 'chemical burn' and severe pain.[8] The stuff you can buy to pour down shower drains to dissolve blockages – that's a potent alkali and, as the label says, you should most definitely avoid contact with your skin. But a more everyday example is likely to be lurking in your kitchen. Some people even use it as a medicine, but often wind up getting the method of administration painfully wrong.

When garlic is cut into, or crushed, it releases a compound called allicin. This is a potent defensive weapon. It can slow the growth of, and even kill, a wide range of microorganisms, including antibiotic-resistant bacteria. (Something it achieves partly by inactivating enzymes that are crucial for generating energy.) Garlic has been used for millennia in medicine, by everyone from the Babylonians to the Vikings, as a treatment for intestinal problems, worms, respiratory infections and much more.[9] There's some modern evidence to support at least a few of these health claims. But while garlic is certainly safe to consume, rubbing it onto your skin is not recommended . . .

In 2018, a doctor reported the case study of a British woman with a fungal infection of her big toe. She had sliced up garlic and applied it raw to her toe for up to four hours a day over four weeks, hoping to kill off the fungus. She might have partially succeeded in that, but at the cost of second-degree chemical burns that left her toe horribly blistered and swollen.[10] This ill-advised home treatment was certainly painful, because it triggered alkali-sensing channels in polymodal nociceptors, but also because it physically damaged cells, which is sensed by nociceptors and leads to pain.

We've looked now at heat and chemicals. The third major class of damage and, so, a cause of pain, is physical – mechanical.

Pinching, crunching, stamping, pricking, breaking . . . mechanical impacts span everything from a needle-pricked thumb to the types of torture methods favoured by the Spanish Inquisition (though they weren't averse to using burning coals, too). Exactly how damaging or potentially damaging forces are sensed is still being investigated. But it's known that when cells are damaged, they leak chemicals that either excite nociceptors, or cause immune responses that make your nociceptors more trigger-happy, or both.[11]

Imagine you've just trapped a finger in a door. When cells are crushed, they release a couple of peptides known as substance P and calcitonin gene-related peptide. These act to expand capillaries nearby, allowing an army of immune cells to flood to the area, and causing swelling in the process. Before long, then, your finger will start to balloon.

Histamine is another compound with a key role in the inflammatory response to injury or infection (and to certain allergens, which is why we take anti-histamines for hay fever). Histamine expands blood vessels. It also makes blood vessel walls more permeable, allowing first-line defence agents, such as the chemicals that help blood-clotting and those that form antibodies, out into the damaged tissue. Like substance P and calcitonin gene-related peptide, it also makes pain neurons more trigger-happy.[12]

Whether the damage is mechanical, chemical or thermal, then, compounds released in the course of the immune response make the region super-sensitive. Inflammation is to damage-detection what a drought is to Australian bushland. One tiny spark, and all hell can break loose. This is why the swollen, red regions around the site of an injury are particularly painful, and why anti-inflammatories relieve pain.

Some types of cell damage can also be itchy. Sunburnt skin, for instance, can itch as well as hurt. But pain and itch feel distinctly different, of course, and they cause you to do very different things to try to ease the sensation (if you cut yourself, you certainly don't feel like scratching the injured site). There is a distinct class of sensory neuron, known by the catchy name of 'MrgprA3+ sensory neurons', that mediate itch.[13] But one of the molecules that they have receptors for, and so that will set them off, is histamine. Sun damage can be itchy, then, because of the local release of histamine in the damaged skin – and an anti-histamine drug will often help with the itch of 'heat rash' on mildly sun-damaged regions.[14]

If something hurts us, it's a prime candidate for hacking by other species that don't especially want us, or other animals, near them. We already know that the chilli pepper evolved to tap into our painful-heat sensing system. But various plants and animals exploit the fact that histamine can trigger both pain and itching. Bee venom, for example, triggers an explosion of histamine – and so stinging pain and itching. The tiny, hollow, defensive, needle-like hairs of the stinging nettle, meanwhile, include histamine in the nasty chemical cocktail that they inject into anything witless enough to get close enough to eat them.

Think about the worst pain you've ever experienced. Was it twisting, squeezing, burning, or stabbing, smashing, bursting, unbearable, pounding, or stiff, ripping, excruciating, dull, agonising, or vicious? Maybe, if you've given birth, for example, it was a case of many of the above.

Take another look at that list, and you'll notice that some of the adjectives we commonly use for pain are discriminative – they describe the type of pain (twisting, stabbing, and so on). Others describe how disturbing it is – how agonising, or excruciating (or, if it's not that bad, how mild or dull it is). As the recent addition to Aristotle's original definition makes plain, we now understand that there are two dimensions to the pain experience. First, there's the discriminative element – where is the pain and precisely what type is it: sharp, tearing, and so on? Then, there's an emotional component: just how disturbing is it – how exhausting or unbearable, or even welcome?

There is no single brain processing hub for signals from damage sensors. But there are various pathways that this information travels along,[15] and also a network of brain regions that are typically activated. This network is typically known as the 'pain matrix'.[16] One important pain pathway carries information about the damage from the spinal cord to the hypothalamus, which orders an automatic increase in heart rate, breathing, sweating and a diversion of blood flow to the muscles: physiological changes that will help you to fight off whatever's damaging you, or to flee.

This damage information is also carried to a region just above: the thalamus, our sensory relay station. From here, it gets sent on to the primary and secondary somatosensory cortices. Processing in these regions allows you to discriminate the specific location and type of pain – if you've just stepped on something, where exactly in the foot is it hurting? And is it broad and angular or spiky? Is it Lego, maybe, or an upturned drawing pin?

Incoming damage signals also travel along another pathway, from the thalamus to the amygdala (like a tireless meerkat sentry, this region is always alert to potential danger), the insular cortex (which receives all inner-sensing information and represents internal bodily states) and the anterior cingulate cortex (which is involved in emotion, as well as impulse control and decision-making). It's activation of this pathway that is thought to generate the *emotional* aspect of pain – the feeling that it's gruelling, or sickening or unbearable, and something that you do not want to be experiencing again.

Some of the evidence for the role of these various regions in generating pain perceptions has come from imaging the brains of healthy volunteers while they are deliberately hurt. Another important line of evidence consists of studies of people who have, on the face of it, bizarre responses to injury.

Christian Keysers, at the University of Amsterdam, specialises in the neuroscience of emotion and empathy. In the course of his research, he's interviewed various people with congenital malformations of the anterior cingulate cortex. These interviews have been enlightening – if also strange. Keysers tells me, 'One of them is a car mechanic and when he cuts himself, he *knows* that he's cut himself, because there's a response from his somatosensory cortex. But it doesn't carry this aversive motivational value of, "Oh shit, this is terrible, I shouldn't do this again."'

'So, in *Star Trek* there's an android with an emotion chip and when he turns it off, he analyses the situation but he doesn't care about it emotionally. That's what these people are like. The mechanic can analyse the fact that he's been cut, but he doesn't cry, it doesn't have the affective baggage of "ouch".'

This extends to their understanding of other people's accidents and injuries. 'You can show them a movie of someone tripping and badly twisting their ankle, and they can analyse the situation and tell you it's probably not good for them – but if you ask them how disturbing it is to watch that, it's not. By losing the affective baggage of your own pain, you lose the salience of other people's pain, too.'

People who live without the emotional side of pain find it harder to empathise with others in pain, and won't learn as readily to avoid doing things that cause them bodily damage. But there's another group of people who find this important lesson even more difficult.

In these rare cases, the problem is usually not in the damage sensors, or regions of the brain involved in processing pain, but somewhere in between. For example, someone with two mutated versions of a gene called SCN9A lacks a sodium channel that is critical for propagating damage signals along nerve fibres. This means that they don't feel any pain. What might sound like an advantage – no pain! – is of course an enormous disadvantage, as they can fail to notice that they're injuring themselves. Or, if they do, it doesn't bother them.

Geoff Woods, a clinical geneticist at the University of Cambridge who specialises in this condition, famously first came across it when he was in Pakistan and was asked to see a boy who was earning money from street performances in which he walked across hot coals or put daggers in his arm, without showing any sign of pain. Before Woods got chance to meet the boy, he jumped from a roof to 'amuse friends', and walked away from it, only to later die from the injury it had caused to his brain.[17]

The journey of damage data from receptors 'up' to the key pain-related regions in the brain is known as the 'ascending pathway'. But this alone does not determine how much a slap or an inflamed knee or even a bullet actually hurts. Signals from the brain 'down' can exert all kinds of influences over how much pain we perceive – and whether it bothers us, or not.[18]

In 1843, the legendary explorer and anti-slavery campaigner David Livingstone set up a missionary station in the beautiful valley of Mabotsa, South Africa. 'Here,' he wrote in his book *Missionary Travels and Researches in South Africa*, 'an occurrence took place concerning which I have frequently been questioned in England, and which, but for the importunities of friends, I meant to have kept in store to tell my children when in my dotage . . .'[19]

A troop of lions had been attacking cows. Livingstone undertook to help local men to kill one or two, in the hope that this would scare off the rest. He did succeed in shooting a male lion – 'I took a good aim at his body and fired both barrels' – but it wasn't dead. While Livingstone was re-loading his gun to have another go, he heard someone shout out:

Starting, and looking half round, I saw the lion just in the act of springing upon me . . . He caught my shoulder as he sprang, and we both came to the ground below together. Growing horribly close to my ear, he shook me as a terrier dog does a rat. The shock produced a stupor . . . it caused a kind of dreaminess, in which there was no sense of pain nor feeling of terror, though [I was] quite conscious of all that was happening. It was like what patients partially under the influence of chloroform describe, who see all the operation but feel not the knife.

The absence of pain, though his bone was crushed 'into splinters' and the lion made no fewer than eleven teeth wounds in his upper arm, was, Livingstone went on to surmise, 'a merciful provision by our benevolent Creator for lessening the pain of death'.

Livingstone's is far from the only story like this. Some battlefield tales also testify to the fact that it's possible to suffer the kind of physical damage that sends damage sensors into a frenzy, but to feel no pain. However, Livingstone's interpretation of the reason why is perhaps overly negative. A more positive one is that, with his life in extreme danger, agonising pain would have stopped him from doing whatever he could to survive. (In the event, another man took a shot at the lion, which then ran to bite this man – an act that was no doubt possible because it was experiencing a similar suppression of pain – before dying from its wounds.)

While the evolutionary purpose of pain is to cause us to remove ourselves from danger and to engage in behaviour conducive to healing – i.e. to rest and refrain from using the damaged body part – and also to teach us a lesson, it's not always helpful to feel pain. The result, as Giandomenico Iannetti, a sensory neuroscientist at University College London, points out, is that we have ways to moderate it. When it comes to acute pain, Iannetti says, 'generally, you feel what it's useful to feel.'

What we feel can be influenced in a number of different ways. We do, of course, produce our own painkillers. Incoming signals from damage sensors trigger the release of a variety of pain-inhibiting compounds, including endorphins and enkephalins (your endogenous opioids, which reduce pain by binding with what's known as the mu-opioid receptor), and also cannabinoids.

A woman in Scotland who was found to have a gene mutation that enhances endogenous cannabinoid signalling hit media headlines in 2019.[20] As well as feeling virtually no pain, she's almost invariably happy, and never feels anxious, though she does suffer from a poor memory . . . It seems that she naturally experiences a state that most people have to consume cannabis to attain.

This important brain-down (rather than body-up) 'descending' pain-killing process inhibits the damage signals coming in from the spinal cord, effectively dampening them.

But how could Livingstone have felt no pain, when spraining an ankle – or giving birth, for that matter – can be agonising?

One probable reason is that his amygdala was screaming DANGER! at his hypothalamus, so increasing the rate and strength of his heart beats. And inner-sensing signals, sent from the heart to the brain with each contraction, inhibit certain types of sensory processing – pain processing, in particular.[21] The faster your heart beat, the deeper the dulling effect on pain perceptions, enabling you to keep on running – or fighting – to save your own life.

Also, Livingstone could have been far too busy focusing his limited conscious attention on the dangers posed to him and on how to get out of the situation to be aware of the damage signals streaming into his brain. This phenomenon is now being exploited to address pain during what are normally excruciating medical procedures. In research at Sheffield Hallam University, for example, two immersive virtual reality games created by game developer Ivan Phelan have been used by people with severe burns while their dressings are being changed, with exactly this in mind.

In one game, players get points for throwing basketballs into hoops; in the other, they must travel around a landscape rounding up sheep into a pen. A young woman called Megan Moxon, who accidentally poured boiling water over her stomach and leg, talks about the difference that playing a game makes: 'Because you're not looking at your dressing while it's getting changed you don't think about it. I had a really bad, white part of my skin and that was so sensitive. But once I was in the VR, I didn't flinch at all. I didn't feel a thing.'[22]

Perhaps because Moxon had been told that playing the game should lessen her pain, a shift in her pain expectations made a physiological difference, too. When we *expect* something to hurt a lot, our brains 'dial up' signals from damage sensors, making it feel more painful.[23] The flipside is that when we expect something not to hurt, the pain isn't generally as bad. Tell someone that a sugar pill is a 'painkiller' and it can trigger the release of their own endogenous opioids, which dampen down signals from damage sensors, and dull pain.[24] Placebos can, then, have observable biological effects.

Because our attitudes to pain influence what we feel, and because our own painkillers can make us feel good, some people experience a sense of pleasure from pain. Self-flagellation, initiation ceremonies, the 'runner's high' resulting from the damage sensors' response to acid produced in hard-worked muscles, as well as, of course, chilli-eating contests – there are all kinds of scenarios in which some of us experience pain *and* pleasure.

But even people who enjoy some pain scenarios don't welcome all the very many varieties. So if you're into S&M, but you've just stubbed your toe against a door, what could you do to turn down the pain?

You could always swear. Because it really is quite effective, confirms a study run at Keele University. Volunteers were asked to plunge their hand into ice-cold water for as long as they could manage. Some were allowed to utter a swear-word (they could choose which-ever they wanted) while they did this, and these people said they felt less pain. These claims were supported by the fact that they managed to keep their hand in the freezing water for an average forty-four seconds longer for men and thirty-seven seconds longer for women. It's not entirely clear why swearing helped, but it might be because it triggers the 'danger' flag, which can dial down pain, the researchers speculated. (Further research has found similar results for Japanese-speakers; this effect isn't, then, restricted to the UK, where swearing in response to pain is much more common and culturally acceptable.)[25]

You might also try to confuse your own brain. Giandomenico Iannetti has led research on this. He applied a hot laser to the back of a group of volunteers' hands and found that when they crossed one arm over the other, their pain was lessened. He thinks that messing with the brain's normal mapping of signals from the right hand to the left side of your world, and vice versa, was responsible.

We all know, though, that if you've just bashed a limb, rubbing can help. This is thought to be because pain and touch signals are integrated in the spinal cord; A-beta fibres, which carry touch messages, connect with the same secondary neuron – the next step in the sensory-receptor-to-brain-transmission pathway – as C fibres. A swamping of touch signals can reduce the damage signals sent on.[26]

Pavel Goldstein has specifically explored the touch of another person. His studies have revealed that while touch from a stranger doesn't mitigate pain from heat applied to the forearm, touch from a loved one does. The reason, it seems, is that when you're with someone whom you're emotionally close to, a state of physiological synchrony develops. Your heart rate, your breathing and even brain-waves begin to fall into step, and this activates the brain's reward network, promoting analgesia. Pain can disrupt this synchrony, Goldstein has found – but touch can re-establish the connection.[27] 'Probably romantic touch is one of the most powerful but not the only possibility for analgesic effect.' However, he adds, 'touch is not a panacea and should be used very carefully.'

Touch is undoubtedly an ancient antidote to pain. But well before the development of general anaesthesia in the mid-nineteenth century, we did, of course, have access to various pain-reducing drugs.

Braziers dating to around 3000 BC containing seeds and the burnt remains of cannabis (which acts on our own cannabinoid receptors) have been found in the Caucasus, and there's evidence that opium from opium poppies was used in Mesopotamia at least a few centuries before that. Archaeologists have even uncovered what looks to be a large-scale drug-production facility at a site in the city of Ebla, not too far from Aleppo, in northwestern Syria, dating from about 4,000 years ago. Traces of poppies (for opium, for pain relief), heliotrope (used to treat viral infections) and chamomile (to reduce inflammation) were all found in large pots at the site.[28] In Ancient Greece, willow bark, which contains salicylic acid, which itself (like a synthetic version, acetylsalicic acid – aspirin) can ease pain and fever, seems to have been regularly used.

Opium hasn't always been taken for its impacts on physical pain, however. Romantic-period poets including Samuel Taylor Coleridge and Thomas de Quincey celebrated its stimulating, creative effects. But Dr Joseph Crawford at Exeter University has analysed the attitudes and experiences of women writers of the same period and concluded that they relied on opium more for its sedative, calming effects, and as a way to cope with distress and depression.[29]

Research confirms that opium and heroin, like our own natural opioids, relieve psychological pain. So, too, in fact, can paracetamol, which affects activity in the anterior cingulate cortex and insula, two regions of the brain that are involved in the emotional response to damage signals and that are also associated with feeling psychological pain as a result of social rejection.

As the US and Canadian team behind a now classic 2010 paper showing that paracetamol can alleviate social pain write, there is 'substantial overlap between social and physical pain'. That's not just in the way the brain processes these types of pain but in the biochemistry of their suppression.[30]

When we talk about rejection by our peer group, or a romantic break-up, as *hurting*, that's exactly what it is. From an evolutionary perspective, there's a really good reason for this. If you were to drive too fast, crash and break ribs, the resulting pain would punish you for doing something that threatened your survival, and make you think twice about doing it again. If, in an ancestral society, in which individuals relied heavily on each other, you were rejected from your social group, your survival would also be at real risk. Stealing from a neighbour or sleeping with someone else's partner, say, are the social equivalents of driving too fast.

Recent research on societies from around the world strongly suggests that the feeling of shame evolved to punish us for breaking the rules of our social group.[31] It functions, then, rather like a psychological version of more moderate physical pain. But social exclusion, which, for our ancestors, would have posed even more of an existential threat than a few broken ribs, *really* hurts. When we're in pain, we're consumed with the desire to be rid of it. And doing anything you could think of to remove the pain of social rejection would have drastically improved your survival chances.

It's worth emphasising, though, that while shame and this kind of social pain, are at root, adaptive, that doesn't mean that they're always helpful or that you can always mend what's socially broken. This can be especially true if your particular group has developed intolerant or skewed social rules that mean that your shame and exclusion result either from something over which you have no control (the way your face or body looks, for example), or from an act or way of life that does not inherently put anyone else at risk.

After all this, you may be wondering – if pain relates to chemicals, mechanical impacts, social threats and heat – why are loud sounds and bright lights so painful? When it comes to bright light, the answer came as recently as 2010, with a study demonstrating that a large proportion of nociceptive neurons that respond to pressure on the surface of the eye also respond to bright light. (At least, this was the case in rats.) In 2015, researchers at Northwestern University published details of what seems to be the ear's own pain system. They found that while the cochlea doesn't contain standard nociceptors, it is home to a set of neurons that are activated only by dangerous noise levels – though whether they are triggered by the death of auditory hair cells (caused by loud sounds) or the sound waves themselves is not yet known.[32]

What all of this research on physical and psychological pain makes abundantly clear is that pain is critical for survival. Well, most of the time. There isn't space in this book to write about chronic pain, or the dangers of addictive prescription painkiller use, both of which have been fully explored elsewhere. But research into the biochemical detail of the sensory processes involved in pain, and its relief, and also the psychological factors that influence it, hold the promise of better methods for helping us *not* to hurt when it just doesn't help.

I I

Gut Feelings

Learn to make better decisions

I generally suffer most on the second day. After that there is no very desperate craving for food. Weakness and mental depression take its place. Great disturbances of digestion divert the desire for food to a longing for relief from pain. Often there is intense headache, with fits of dizziness, or slight delirium. Complete exhaustion and a feeling of isolation from earth mark the final stages of the ordeal. Recovery is often protracted, and entire recovery of normal health is sometimes discouragingly slow.

This is Emmeline Pankhurst, a leader of the suffragette movement in the UK, writing in 1912 about her periods of self-inflicted starvation during imprisonments for violent protests.[1] For Pankhurst, as for fellow militants who sought equal voting rights for women, hunger-striking was a way to draw attention to the government's refusal to recognise the movement's incarcerated protesters as political, rather than criminal, prisoners.

The first documented case had taken place only three years previously, when the artist and suffragette Marion Wallace-Dunlop was convicted of maliciously damaging the stonework of the House of Commons (she had stencilled an extract from the English Bill of Rights on a wall in St Stephen's Hall), and sent to Holloway Prison with the classification of criminal offender. Her refusal to eat 'until this matter is settled to my satisfaction' lasted ninety-one hours. At this point she was released, for fear she would die.[2] Hunger-striking was quickly adopted by fellow suffragettes. It was a powerful and emotive non-violent political tool, one that would later be taken up by, among others, the IRA, Mahatma Gandhi and Nelson Mandela, during his captivity on Robben Island.

Hunger is of course the conscious feeling that we want food. It evolved to motivate us to act on sensory signals indicating that the body needs more fuel. To go on hunger strike – to reject food – means deliberately refusing to act on one of the most compelling drives that anyone can experience. Christabel Pankhurst, Emmeline's daughter and a fellow leader of the Women's Social and Political Movement, observed that the practice marked the 'triumph of the spiritual over the physical'.[3]

What hunger evolved to do, and the contexts in which we often experience it today, are two different things. But it certainly can signal that we need more key nutrients, such as carbohydrates, to keep ourselves going. That process relies on a system for sensing stocks in the body – a fundamental system that is common to plants as well as animals. Whether you're talking about a rose bush in your garden, a slug, or a goose having to be force-fed grain for the foie gras industry, all can sense when they require a top-up of vital nutrients and water, and when they have too much.[4]

For plants, one key nutrient sensor was identified only recently. It responds specifically to phosphorus, which they need for normal growth. These sensors tell the roots when to suck up more phosphate (the source of phosphorus from the soil), and when to stop. And just as hunger drives us, along with other animals, to actively search out food, phosphorus starvation has this effect on plants: it stimulates roots to grow out sideways, to spread through the top layer of the soil where phosphate tends to accumulate and so is more likely to be found.[5]

For us humans the sensory signals that relate to food consumption and digestion are a lot more complex, of course. They're also far-reaching, affecting our emotions, thoughts and behaviour, exerting potentially profound effects on our lives.

In 1912, a year in which suffragette violence, imprisonment and hunger-striking escalated, the American physiologist Walter Cannon wrote that hunger is 'characterised by highly disagreeable pangs which result from strong contractions of the empty stomach – pangs which disappear when food is taken'.[6] Today, 820 million people globally are under-nourished, meaning that their calorie intake is below the minimum that their body needs.[7] For them, hunger pangs will be a familiar experience. But in many countries, food is over-plentiful, and we now know that the causes of perceptions of hunger are more complicated: they are driven by a range of signals from within the body, as well as, of course, by habit – as anyone who's ever suddenly felt peckish after walking into the kitchen knows well.

In fact, think for a moment about the last time you ate something. What prompted it? Did you actually feel *hungry*? Or was it just 'time' for breakfast? Or were you tired, and hoping that a snack would perk you up? Surely you can also recall a time when you went out to dinner not feeling hungry at all, but at the first whiff of something delicious, you suddenly discovered an appetite, after all.

Feelings of hunger – a mental state – can certainly happen even when our bodies aren't telling us that we're running on empty. And when we do start eating, all kinds of external factors can influence when we feel that we've had 'enough'. Even something as basic as the size of your plate has an impact. Put a portion of food on a big plate and you'll perceive it as being smaller than exactly the same portion arranged on a smaller plate. (It's not only humans who fall for this illusion; even some reptiles do, too.)[8]

Being able to focus in on our stomach signals, then, could be hugely beneficial to those trying to control their weight. Physical, mechanoreceptive signals about the degree of distension of your stomach and your intestines affect our perceptions of fullness and hunger,[9] and in a sensorily-confusing world of plenty, they could be the most reliable guide to how in need of food you truly are.

Historically, researchers who wanted to probe our perceptions of stomach fullness used unpleasant techniques involving inserting balloons down through the oesophagus, then inflating them. A newer 'more participant-friendly' (as one research paper puts it) technique requires volunteers to drink water to the point that they feel their stomach is 'comfortably' full, and then more, until they feel it is 'maximally' full.[10] These perceptions are driven by signals from mechanoreceptive channels in the endings of sensory neurons in the stomach. However, recent work has shown that the sensitivity of these receptors can be affected by certain hormones – specifically, by hormones released by gut cells equipped with taste receptors.[11]

As we know from Chapter 4, these taste-like cells detect levels of various nutrients in the food that is being digested. So it makes sense that their signals should affect how full we feel. It's thought that if you're eating a meal packed with lots of nutrients, mechano-receptors will signal 'Full!' when the stomach is less distended than if you're eating a meal with a relatively meagre nutrient content. This makes perfect sense. Imagine a bowl of soup dense with chunks of potato and chicken, and another holding a watery broth. You'll have to eat more of the broth to get vital nutrients in. So the volume of a meal alone should not be key to how full you feel. Allowing signals from one key type of eating-related sensing ('taste' outside the tongue) to influence the action of another (physical stomach stretch), helps to fine-tune our perceptions of when we should lay our cutlery down.

To return now to those water-drinking studies – by testing and re-testing people, it's possible to establish how consistently one person feels 'comfortably' or 'maximally' full after drinking the same amount of water. People who, across many tests, reliably report that the same volumes of water make them feel 'comfortably' or 'maximally' full are deemed to have better 'gastric interoceptive accuracy' – or 'stomach-sensing'. Studies show that a person's stomach-sensing accuracy correlates with their heart beat awareness.[12] Ability at these two types of inner-sensing is related.

There's also some evidence that people with the eating disorders bulimia or anorexia aren't as good at stomach-sensing; they're just not as accurate as other people at perceiving when their stomach is actually full, or how full it really is. In fact, they seem to have more generalised difficulties with monitoring inner bodily signals (as the correlated stomach/heart beat results would suggest).[13] Whether this could be part of the cause, or, alternatively, partly a result, of their eating disorder is not yet known, but psychologist Rebecca Brewer at Royal Holloway, University of London, is working on it. She's hoping to run longitudinal studies, following a group of people from the age of ten and checking in on them every eighteen months or so, measuring their inner-sensing ability each time, and noting any change in this, and also the emergence of any eating disorders.

If poor stomach-sensing is a part of the cause of an eating

disorder (and to be clear, no one is suggesting that it might explain a complex disorder like anorexia entirely), why might some people be better at it than others? Might our genes influence our stomach-sensing capability? No one has looked explicitly at this. But as a spectrum of linked sensitivity has been found for other senses that rely on mechanoreceptors – as Gary Lewin has found for touch, hearing and also blood pressure sensing – it would seem at least plausible.

For people who don't eat too little, but rather too much, in theory, learning to pay closer attention to the physical signals of stomach fullness could help. As Karyn Gunnet Shoval, formerly an instructor at the Yale Stress Center at Yale University, now a health coach, explains, it can be useful to learn to identify how much you really need to eat, based on physical stomach sensations, on a scale of 1 to 10.

Strategies such as drinking a cup of water, then focusing on how your perceptions of stomach fullness have changed, can help with this. By learning to home in on these signals, Gunnet Shoval says, we can then use them to estimate our true hunger. We can learn that, even if it's 'time' for dinner or habit is driving us to eat something, if it's a '5' on the scale, say (which denotes 'I'm hungry but not so hungry that if I don't eat something right now it'll be a big problem'), we don't actually need to eat anything; while if it's a '3', we need to eat, but we need to consume less than if it's a '1'. 'It's about, "What should I be eating to help me feel better?", without going to the extreme,' she explains.

An alternative strategy, if you do want to try to eat less, might be simply to *imagine* that you're stuffed. In a recent experiment at the University of Utrecht, a group of people spent a minute imagining being either hungry or full. Then they were asked to choose between a variety of food options, including varying portion sizes of popcorn, chocolate ice cream and crisps, as a 'reward'. Those who'd imagined being full chose smaller portion sizes than those who'd imagined being hungry. This implies, the researchers observe, that just mentally simulating visceral states, such as feeling full, can affect real choices with immediate consequences.[14]

If you're actually really in need of food, though, just trying to imagine feeling full of course won't cut it. And though Emmeline Pankhurst and her fellow hunger-strikers were able to use sheer force of will to starve themselves for days in the interests of their cause, for the rest of us, hunger can bring out our less noble sides, driving us to do whatever's necessary to get our hands on some food. That, anyway, is the explanation for 'hanger', a term that combines hunger and anger, which can entail grumpiness in an adult who's trying to behave and a full-scale tantrum in a less-regulated small child.[15]

While many of us are probably familiar with hanger, at least to some extent, the psychological impacts of sensory signals that relate to food don't end here. In everyday conversation, we often talk about 'gut feelings': You're not sure whether to buy House A or House B? Go with your gut. These feelings tend to feed into big decisions – the kind that are characterised less by a conscious weighing up of the pros and cons, and more by an unconscious swing towards one option over another.

Just because the swing is unconscious doesn't mean of course that it's uninformed by fact. Implicit knowledge, or learning, consists of associations, or patterns, that your brain has learned at one level, but that you haven't quite clocked. This is an evolutionarily primitive type of learning, and though there is a spectrum of ability in the general population, it does not relate to IQ.[16] It's possible to get a Mensa-level score in an IQ test but be an implicit-learning dunce, or vice versa. When you look more closely at the results of lab studies that reveal implicit learning in action, it's immediately clear why.

Many of these studies have involved tracking people's physiological states while they play a gambling game with rules that they have not been made privy to. To do well at this game, then, a player has to try to fathom the patterns of action and effect – the rules – as they go. However, they are set to be so complex that it's very hard to work them out logically. Still, given experience, some players, at least, will start to show that they have learned them, or at least got sufficiently to grips with them that they can make wise decisions – choosing to gamble when the chance of winning is in fact high, and holding back when that chance is low. When researchers ask these participants what the rules actually are, though, they'll generally say that they have no idea. So *how* have they learned them?

The answer lies in their physiological states.[17] While playing the game, some of those states (such as their heart rate and the sweatiness of their skin) are measured. The studies show that these states become subtly different in situations when it would be wise to gamble, compared with those in which it would be a bad idea. The brain loves to win. And the amygdala is quick to learn 'winning' situations (whether that's in a gambling game or something else that affects your ability to survive and thrive) and 'losing' ones.[18]

When it spots a threat, it triggers Walter Cannon's flight or fight response, instructing the hypothalamus to increase your heart rate and sweat level, and inducing other bodily changes. If the threat is crystal clear, that response is strong. If it's fuzzy and faint (triggered, say, by hints that you might be holding a losing hand), it's weak. But you can still detect the fuzzy, faint sensory signals of a weak physiological response to a threat, without becoming conscious of what that threat actually is. Your bodily signals can, then, nudge you towards making an unconscious, intuitive (rather than a logically thought-out) correct decision. People who are in tune with their body-state signals should, then, be better at implicit learning. And, since inner-sensing ability and awareness do not relate to IQ, neither should someone's ability to engage in this type of learning.

Given that we aren't equally in tune with these inner signals, Hugo Critchley and Sarah Garfinkel at the University of Sussex wondered whether this might affect real-world decision-making and success. To explore this, they wanted to find a group of people who have to assimilate very large amounts of information rapidly, spot patterns in complex data sets, and make quick and highly risky decisions. They hit on stock market traders. These people have to do all these things, and good decisions are rewarded with stacks of cash and still having a job, while poor decisions could put you out on the street. In theory, then, being better at bodily sensing might be extremely valuable.

With colleagues, Critchley and Garfinkel studied traders on the London Stock Market. Their results, published in 2016, were striking. First, the traders were better at heart-beat-sensing than a matched group of people with different jobs. This indeed suggested that being good at inner-sensing helps with trading success. But the team also found that an individual trader's inner-sensing ability predicted how much money this person made – and how long they survived on the floor. Within this elevated super-spectrum of inner-sensing ability, the ones with better scores were more successful.[19]

'Traders in the financial world often speak of the importance of gut feelings for choosing profitable trades,' the team writes. 'Our findings suggest that the gut feelings informing this decision are more than the mythical entities of financial lore – they are real physiological signals, valuable ones at that.'

Registering a slight change in your heart rate isn't itself a 'gut' feeling. But, as we know, heart-rate-sensing is widely used as an indicator of general inner-sensing ability. In a threatening situation, blood is diverted away from the intestines, while adrenaline makes the involuntary smooth muscle of the intestines relax, and stretch receptors in these muscles detect this. Your brain registers these changes, and come to spot the situations in which they tend to happen, without you ever consciously realising it, but giving you the feeling of having a 'hunch'. This seems to be the case for the more inner-sensitive financial traders. And it's why when someone is unsure what to do, they're often advised: 'trust your gut'.

If 'trust your gut' is a scientifically backed bit of old wisdom, then what about the claim that we should all be drinking two litres of water a day? Countless articles tout drinking two litres of water as the solution to weight loss, fewer wrinkles, better concentration and more. But how true is this? To get to the bottom of it, we need first to understand how thirst works.

We need to drink often because, unlike a camel, say, we haven't evolved extraordinary methods of conserving water inside us. As the famous rule of thumb goes, you can survive about three weeks without food but only three days without water. (The actual number of days will, depend, of course, on how much water you're losing in sweat and breath, and whether you're also vomiting or suffering from diarrhoea, both of which make that fatal timeframe shrink.)

Yet go below the surface of your body and there's water, water everywhere . . . Back in 1945, a pioneering group of physiologists at the University of Illinois revealed just how much – and where.

The team took on the task of analysing the chemical make-up of the body of a thirty-five-year-old man who had recently died. He weighed seventy kilograms and was 183 centimetres tall. That's lighter and also a little taller than the average American man alive today, but his vital element statistics are still quoted, because the team left no organ unturned in their quest to break him down into his material constituents. The human body, they concluded, was, among other things, made up of 14.39 per cent protein, 1.596 per cent calcium, 0.771 per cent phosphorus – and 67.85 per cent 'moisture'.

Their grisly table of results reveals that water accounted for almost 74 per cent of the weight of his heart, and the same for his brain and spinal cord. Though his kidneys topped the water chart at 79.47 per cent, almost a third of the weight of his bones was water. Even for his teeth, the figure was 5 per cent.[20]

More recent research suggests that the overall figure of 67 per cent of an adult's weight as water is at the high end of normal. But still, if you could somehow be sucked dry, you would certainly lose around half your body weight (a little less if you're particularly muscly and significantly more if you're carrying plenty of fat).[21]

We need water for all kinds of essential functions: for blood, to transport oxygen around the body; to get rid of waste in urine and faeces; to cushion the brain from impacts; for sweat, to regulate our body temperature – and so much more. It's no wonder, then, that we are exquisitely sensitive to changes in the concentration of dissolved chemicals and minerals, and therefore the relative water content, of our blood. A higher concentration indicates dehydration – and that's an immediate problem, because it can mean that cells aren't maintaining the volume required for them to do their jobs properly.

For more than seventy years, physiologists have known that the brain not only regulates the water content of bodily fluids, but directly senses it. There are cells in regions of the hypothalamus (the master-controller of homeostasis) that directly detect the concentration of dissolved particles and, especially, sodium (from salts) in your blood.[22] If this concentration strays just the tiniest bit outside an extraordinarily tight range of acceptability, measures to adjust this immediately kick in. One: the kidneys are instructed to lose less water in urine, which means more is kept in the blood. Two: you feel thirsty, which prompts you to drink.

For a long time, it was thought that this brain-sensing of blood hydration was the whole story of thirst. But this niggled some physiologists. After all, it takes ten minutes after you've downed a drink for your blood water level to change. And yet we're all familiar with that sensation of taking a long, cold drink and feeling that it's immediately quenched our thirst. In 2019, a team at the University of California, San Francisco reported why. In this study, they used mice as models, and found that when a mouse starts to drink, sensory signals from the mouth and throat temporarily shut down thirst neurons in the hypothalamus. This 'fast' signal, which is especially stimulated by cold drinks, seems to track the volume of liquid that's being consumed.

But even this is not the whole story. Further work on mice showed that this 'thirst neuron switch' is also triggered by salty water – but not for long. Something clearly registers that this is not a hydrating fluid and reverses the 'thirst quenched' decision. And, also in 2019, the team reported what that something was: a second level of sensors, probably in the beginning of the small intestine, that can detect whether, and to what extent, a drink (or some food) will actually be hydrating. Thanks to signals sent along the vagus nerve to the brain, these sensors can then update and refine perceptions of thirst.[23]

These researchers even studied the mouse brain's response to dehydration and drinking – and also hunger and eating – right down to the level of individual neurons. 'This is the first time we've been able to watch in real time as single neurons integrate signals from different parts of the body to control a behaviour like drinking,' noted Zachary Knight, the neuroscientist directing the research. 'This opens the door to studying how all these signals interact, such as how stress or body temperature influences thirst and appetite.'[24]

This takes us to the question of how much water we really *need*.

The origins of the two litres/eight glasses a day advice are a little obscure. But some researchers have traced it to US National Research

Council guidelines, published all the way back in 1945, the same year that the Illinois team analysed their cadaver, which state that adults should 'consume' one millilitre of water per calorie of food. That would indeed equate to about two litres per day for women and two and a half litres per day for men. However, note the word 'consume'.

Much of the food that we eat contains water, and that water counts. Naturally, how much exercise you do and the climate you're in will affect your requirements, too. But for those of us who, like me, live in a temperate climate and spend too much time sitting down, it's thought that drinking more like one litre of fluid each day is all that's necessary. And even that doesn't have to come in the form of pure water. Tea and coffee are often held up to be dehydrating, but that's wrong; they, too, contribute to your total water intake, according to research by Heinz Valtin, at Dartmouth Medical School, USA.[25]

What about the claim that we become notably dehydrated before actually feeling thirsty, so should engage in pre-emptive drinking?

It would seem extremely odd for any animal, us included, to have evolved such a skewed sense of thirst. Water is so vital, and its level in our blood is kept within such extraordinarily tight limits, it would be very strange indeed if we needed it – even a little of it – and didn't feel thirst. The only real caveat to that is that it is possible to become so caught up in what we're doing that we neglect the onset of thirst signals. Kids in a playground are a classic example. Why stop playing to go to the boring water fountain, until you're *desperate*? And if you know you're going to be in a situation where water won't be freely available (such as an exam room), it's a good idea to take some with you.

Overall, when it comes to staying hydrated, the sensible thing, it seems, is to go with how you feel. And if you are trying to drink, and it isn't slipping easily down your throat, that's probably because you're over-hydrated, according to research at the University of Melbourne. The team studied people who'd either recently drunk a lot of water or who were mildly dehydrated, and who were then given a glass of water to sip. The participants were asked to report on how much effort it took to swallow one sip. The dehydrated group gave an average rating of 1 out of 10, indicating the least effort. The other group found it much harder; their ratings averaged nearly 5.[26]

This is evidence of yet another indicator, beyond our level of thirst, as to whether we actually need water or not, the team says. It's also further evidence that we're very good at knowing when and how much we need to drink. 'Just drink according to thirst rather than an elaborate schedule,' advised team member Michael Farrell.[27]

We're also, of course, very good at knowing when we need to deal with the final practical result of drinking (and eating) and go to the loo. Stretch receptors in the bladder and the rectum are responsible for this.[28]

Though our water-sensing and regulation system might be highly sensitive, there is an entire class of people who are often dehydrated: older people. With age, we all tend to get worse at inner-sensing, notes Geoff Bird at the University of Oxford. In fact, he says, one of the most obvious signs of these deficits in older people is dehydration. Since age blunts our instinctive sense of when we need more water, it can be hard to get elderly people to drink enough.

This deterioration may happen at the level of the receptors themselves: they may thin out or become less sensitive with age. 'It could also be to do with the transmission of that information to the brain; its representation in the brain gets more "noisy",' says Bird. And as we'll find in Chapter 14, faltering inner-sensing could be a major contributor to impacts on older people that go way beyond the physical.

For the rest of us, understanding the senses that relate to thirst and hunger can clearly help us eat and drink more healthily. So much dietary advice relates to our food choices – avoid pizza, cut out cake, and so on – but if we consider what our *senses* are telling us, we can learn to trust the more helpful honest messengers (such as stomach and gut stretch receptors) and try, at least, to put the urgings of others (ooh, that sweetness is so nice!) in their place.

Part Three

A Symphony of Sensing

In the first two parts of this book, I've looked at our senses in isolation or in small, tight-knit groups. This is necessary for exploring just what our vast repertoire of senses does for us. However, our senses rarely operate in isolation, but rather come together, in a kind of symphony of sensing – and this can elevate our perceptions from basic sensory observations (that fruit is red, it's a hot day) to a sophisticated appreciation of ourselves, other people and the wider world.

When we talk about having a 'sense' of direction or 'sensing' someone else's distress, for example, we're referring to experiences that fundamentally depend on input from a suite of senses. In this part of the book, I'll explain how both happen.

I'll also look at how patterns of variation in many senses mean that certain groups of people have sometimes radically different experiences of the world. For a start, sex makes a difference: as a group, women don't sense in the same way as men.[1] But we also all know men or women who are incredible navigators, say, or who are hyper-emotional, or who instantly intuit how others are feeling. We also know examples of the reverse: terrible navigators, or emotionally 'flat' types, or people who seem to have no clue about how others are feeling. These are not minor differences in the human experience. Understand how they are underpinned by our sensory 'settings' and the promise is a rich, new insight into your parents and children, your partner and friends and yourself.

12

A Sense of Direction

Why do I always get lost?

How is that some people navigate places with ease, while others get lost at the first missed turn?

I belong to the latter group, unfortunately. My life has never been at real risk because of it – but only because I have consciously never put myself in the kinds of situations in which that might occur. But when I first lived in London, in the days before even mobile phones, my A–Z street directory was as precious as my purse. Now, if I'm walking or driving to a grand total of about five destinations near my house, I manage without sat nav. For anything else, I *always* have my phone, with my plethora of map apps. There's only one place in the world that I've ever felt freed from the constant background fear of getting desperately, hopelessly lost, and I'm sure it's part of the reason I love it. New York City, with your fabulously numbered streets – you made me feel something close to directionally able.

For people like me, a poor sense of direction can feel fundamental and inevitable. Like having brown eyes, say, or long arms. However, our ability to navigate relies on a suite of senses, including some of the senses that we met in Part Two. And as we already know, our senses can be trained. Understand how they come together to allow you to find your way around (and the barriers to this), and you can, then, start to appreciate not only why there's a spectrum of ability, but also how, no matter how good you are, you can help yourself to get better.

One important insight into variations in our navigational ability is this: it bears no relation to general intelligence.

Dan Montello, a cognitive geographer at the University of California, Santa Barbara, supervised a now classic 2005 study that documented this fact. Twenty-four student volunteers were driven individually along a few routes in an unfamiliar residential neighbourhood with lots of hills and winding roads. Then they were asked questions about the spatial layout of where they'd just been. For example, they had to point from one landmark in the direction of another (which they could not see) and sketch a map of the area. This process was repeated once a week, for ten weeks, using a variety of routes past the same landmarks, and all through the same neighbourhood.

Huge differences in the students' performances emerged. Though some gradually improved over the ten weekly sessions, most either 'got it' within a single session – or simply never did; they started out bad, and they didn't get better. Montello dubs this group 'baggers', for people who couldn't navigate their way out of a paper bag . . . Remember, these were all students who'd been accepted into a highly prestigious college. Their memories for facts, and their overall smarts, all had to be well above average. But while some were quickly able to develop a mental map of this neighbourhood, others floundered.[1]

Mary Hegarty, a colleague of Montello's, as head of the Spatial Thinking Lab at UCSB, has led the development of what's known as the Santa Barbara Sense of Direction Scale.[2] It asks you to indicate to what extent you agree or disagree with statements like, 'I am very good at giving directions', 'I have a poor memory for where I left things' or 'I very easily get lost in a new city'. She's found that people's scores on this scale correlate well with results on real-world studies like this one of Montello's, and also with lab-based tests of navigational ability.

Lab tests typically involve VR environments. Generally, they allow for the use of only one type of sensory information: visual. Certainly, when navigating, we do rely heavily on vision, and particularly our brain's ability to monitor what's called 'optic flow'. This is the pattern of the *apparent* motion of objects as *you* move. But blind people can also learn to find their way around – to build up an understanding of where to turn at which corner, and also a map-like mental model of where various landmarks are in relation to each other.

These two strategies (route-based and mental mapping) are both important for navigation. The route-based type relies on remembering landmarks and turnings on a particular route. If I want to get from my house into the city centre, say, I know I have to turn left out of my street, cross the roundabout, veer right at the odd junction that looks as though it's taking me the wrong way into a one-way street, but isn't, then turn left. This type of strategy works pretty well on regular, well-used journeys but it's inflexible. What if there are roadworks at the roundabout, and I can't go straight ahead?

Then I'll have to plot a new route in my head (or switch on the sat nav). This kind of navigation relies on 'mental mapping' of the environment – the type that Montello tested with his students. Finding your way using mental mapping is generally considered to be a superior approach, because it's flexible and allows you to work out shortcuts. But it is more cognitively demanding. Good navigators, says Mary Hegarty, will automatically select the best strategy for the job.

Since blind people have no problem with both strategies, senses other than vision clearly contribute to your ability to find your way. In fact, body-mapping signals, which allow for 'muscle memories' of the movement of your body through time, and especially vestibular signals, which help you to understand which way you're facing, and how fast you're moving, are all important. If you're riding in a car, proprioception won't help. But vestibular signalling certainly will.

What we know about how the brain assembles all this sensory information into useable representations of your environment, which underpin your sense of direction, comes mostly from studies of the activity of single neurons in animals. And a good chunk of that has been done in labs at University College London.

It's not always easy for the Nobel Assembly to convey to the public just why an award-winning development is so deserving but in the case of the 2014 Prize in Physiology or Medicine, they had no trouble: 'The discoveries of John O'Keefe, May-Britt Moser and Edvard Moser have solved a problem that has occupied philosophers and scientists for centuries – how does the brain create a map of the space surrounding us and how can we navigate our way through a complex environment?'

How do we know where we are? How can we find the way from one place to another? And how can we store this information in such a way that we can immediately find the way the next time we trace the same path? As the Nobel Assembly commented, these were desperately important questions that had long gone unanswered. What this trio of scientists had discovered was a positioning system, a kind of 'inner GPS' in the brain that makes it possible for us to orient ourselves in space.

John O'Keefe won his half of the prize for his discovery, in 1971, of 'place' cells in the hippocampus, a brain region important for memory.[3] While using miniaturised electrodes to monitor the activity of single neurons in the brains of rats as they wandered through an enclosure, O'Keefe noticed that some were always active when the rat was in a specific spot in the enclosure, but not when it was elsewhere. But others were active when it was in a different location – and so on. He concluded that memories of place cell activity could effectively function as a map of an environment.

The married Norwegian neuroscientists Edvard and May-Britt Moser, whose post-doc work was supervised by O'Keefe, discovered 'grid' cells in the entorhinal cortex, a region next to the hippocampus, in 2005.[4] The Mosers were also monitoring the activity of individual neurons, and they identified some that, unlike place cells, didn't become active at single locations, but at several locations. These cells fire in groups, each mapping onto a discrete hexagonal region of ground as an animal moves across it. By essentially unfurling a grid map over a two-dimensional space as it goes, the rat obtains precise information about the distance between objects within it, including itself.

Place cells and grid cells have both been found in human brains, as have a variety of other neurons specialised for navigation.[5] For example, head direction cells, which 'everybody forgets about because they didn't win the Nobel Prize,' notes Kate Jeffery, director of the Institute of Behavioural Neuroscience at University College London. (Jeffrey also had O'Keefe as a post-doc supervisor.) 'But over the past 30 years,' she says, 'interest in them has really grown.' In fact, head direction cells encode the orientation of your head, providing a reference point for grid and place cells.[6]

Border cells, meanwhile, fire when you get close to a boundary, such as a wall, while spatial view cells become active when you look at a place, even if you don't actually go there. These spatial view cells, which rats do not have, allow us and other primates to 'use vision in the long range,' as Jeffery puts it. It lets us 'use our eyes a bit like a really long hand'.

Visual signals clearly feed into the inner GPS, especially in an unfamiliar environment.[7] But there's only so much information they can give. How do you know if you're moving, or things around you are moving? You must have had that experience of sitting on a stationary train, with another train waiting beside you, then realising that one has started to pull away – but not being entirely sure which . . . until your brain registers that your vestibular system isn't sending motion signals, so it can't be you.

Signals from the horizontal semicircular canal, in particular, feed into head direction cells. They tell you which way you're facing, and help you to understand where you've been and which way you're going. This canal is, therefore, a key player in navigation.

There is also a sense that can permit quite extraordinary, unerringly accurate journeys over many thousands of miles. Birds do it, bees do it . . . and though it's highly controversial, a few researchers speculate that we may possibly do it, too. This sense is magnetoception – the detection of a magnetic field. For the purposes of navigation, that means Earth's.

A few years back, while holidaying up in the Scottish Highlands, I, along with a hungry common seal, watched Atlantic salmon swimming in through the little estuary of the River Forss, and on, into fresh water. It wasn't easy for them. They were struggling through the shallows of a river depleted by a long, dry summer. Still, this was nothing next to the journey they'd just made. Those salmon would have left their native pool up river a few years before. After swimming off to feed as far away as the waters around Greenland, they were now on their way back to that same stretch of the river, to mate.

Sea turtles do something similar, and many birds and butterflies also make vast migrations. While you (well, I) might get lost while wandering out of a hotel in search of breakfast, in autumn, monarch butterflies manage to fly up to 3,000 miles south from northeastern USA to winter in Mexico.

Precisely how monarch butterflies achieve this – and how Atlantic salmon and sea turtles, or homing pigeons, or migrating Arctic terns – perform their own navigational feats isn't perfectly understood.* But characteristics of the Earth's magnetic field do vary, according to latitude. And there are various ways that animals might sense and track these variations.[8]

In 2012, German researchers found magnetic-field-detecting cells in salmon. These cells, which were taken from tissue extracted from the nose, contained microscopic clusters of crystals of magnetite – a highly magnetic iron oxide. It's thought that they act like tiny compass needles, and when they twist in an attempt to align with Earth's magnetic field, they trigger mechanoreceptors.[9]

Variations in the field might be used, then, as reference points. In theory, a newly-hatched salmon alevin may imprint an element of the local magnetic field – perhaps its intensity. To locate the general area again later in life, it would have to find the coastline, then swim north or south along it to find its river's estuary.[10]

Magnetite has also been found in the beaks of homing pigeons, the brains of sea turtles, and the abdomens of honeybees.[11] And it's been spotted in tiny quantities in the human brain, which has led some researchers to wonder whether we might be able to use it to sense Earth's magnetic field – but, as yet, there is no evidence that we can.

Migratory birds – along with monarch butterflies, and the fruit fly *Drosophila* – also possess a different type of magneto-sensor, which is located in their retinas. It's a protein called cryptochrome, which responds to magnetic fields in the presence of light.

A version of this protein has been found in our retinas, as well as in the retinas of some other mammals.[12] And when fruit flies were engineered to produce the human version, rather than their own, they were still capable of aligning themselves to magnetic fields. (Fruit flies that were bred to be incapable of producing either their own or the human form could not.)[13] Could *we* perhaps

---

* In fact, ornithologists tell me, in the case of birds, certainly, there's a not entirely un-acrimonious race on to crack this, since whoever manages it will gain academic stardom.

use cryptochrome to sense the Earth's magnetic field, to help us to navigate? We do seem to have the hardware. But while other scientists don't dispute the fruit fly findings, there is certainly scepticism that we can actually use the protein for this purpose ourselves. Perhaps, it's suggested, if we do use cryptochrome, since its stimulation also relies on light being present, signals might be feeding into our sense of the time of day – into our body clock. Light intensity is an excellent indicator of the time of day, and the season, but there are also seasonal and diurnal patterns in Earth's magnetic field, and perhaps we use this information to inform our circadian rhythms, too.[14]

In March, 2019, though, Joe Kirschvink at the California Institute of Technology, a long-time advocate of the possibility of human magnetoception, did publish a paper, with his colleagues, reporting that the human brain can unconsciously respond to Earth's magnetic field – or at least, some human brains can, because not everyone in the study group responded during the magnetic field manipulations.

The team built a cube with walls that blocked any electromagnetic radiation from getting in. Each volunteer in turn sat inside it, in darkness, wearing an EEG cap. The cap allowed the researchers to monitor electrical activity in their subjects' brains, and to watch how it changed as they manipulated the magnetic field inside the cube. When the field was oriented downwards and moving anti-clockwise, there was a response – a dip in the amplitude of alpha waves. The team interpreted this as indicating that the brain has realised something needs attending to: that the position of the magnetic field (which would typically be Earth's) is moving while the person isn't.[15]

Not everyone responded to the same degree, but some people showed substantial changes in alpha wave amplitude. In one person, it temporarily dropped by 60 per cent. What sensor might be involved? Since the studies were conducted in darkness, magnetite might seem like the best contender. But as yet, no one knows.

Overall, though, whether or not humans can sense and actually use Earth's magnetic field remains controversial. But what is undisputed is that we certainly *can* use senses other than vision, proprioception and those provided by the vestibular system to help us to navigate.

Though Atlantic salmon presumably use magnetoception to get to the vicinity of their native river, there's good evidence that it's their acute sense of smell that guides them back to the exact pool where they hatched. Other animals, too, use smell for navigation – us included. In fact, we're capable of sniffing our way, blindfolded, back towards a location that we've only smelled once before.[16]

This finding, from psychologist Lucia Jacobs, at the University of California, Berkeley, fits nicely with recent animal work identifying cells in the hippocampus that respond to non-spatial features of an environment – such as odours and textures – and that fire like place cells when an animal moves around in space.[17] It's thought that this system allows for memories of 'what happened where'. For my dog, it might be a case of, 'Yes, it was under that tree over there that I found that bit of discarded kebab! That's a spot worth returning to!' For you, it might be: 'Yes, that incredible aroma of coffee! – it was round the corner, and down the alley . . .'

In 2018, a Canadian team reported that people who are better at remembering smells (in this study, the volunteers had to identify a range of odours, including basil and strawberry) are better at navigating virtual environments. The team went on to find that people with damage to the medial orbitofrontal cortex, which is important for processing odours, not only had trouble identifying – and so remembering – smells, but also had problems with spatial memory.[18]

Our distinctive, pyramid-shaped nose appeared on the Homo evolutionary scene with *H. erectus*. And Jacobs theorises that it evolved to be that way to facilitate long-distance navigation. *H. erectus* evolved in a highly unpredictable climate, she notes, when forest habitats were changing to grasslands. These changes in climate and habitat encouraged the evolution of traits that improved the ability to move around on two legs, which allowed archaic humans to travel further to find food and other resources. (*H. erectus* was the first known hominin to migrate out of Africa.)[19] But finding food is one thing. For your family to survive, you have to bring it home – and that required decent distance navigation. 'Olfaction is like this background fabric to our world that we might not be conscious of, but we are using it to stay oriented,' Jacobs says. 'We may not see a eucalyptus grove as we pass it at night, but our brain is encoding the smells and creating a map.' [20]

Early sailors and navigators reported the use of smell to navigate. Sagas describing the journeys of Viking mariners recount, for example, how they used their senses to sail out of sight of land – and return. They would watch whales feeding in particular currents; listen for bird calls and the sound of waves breaking on rocks; taste the sea, for hints of fresh water flowing in from rivers; and smell land on the breeze. Early Polynesian sailors, who travelled up to several hundred miles between islands, and back, reportedly used the smell of land, too, to help them to reach their destinations.

In fact, modern Pacific Island sailors who rely on traditional navigational techniques report extraordinary uses of their senses. [21] The New Zealand-born sailor David Lewis details many methods in his wonderful book, *The Voyaging Stars: Secrets of the Pacific Island Navigators*. Among the many memorable stories is one concerning a man called Tevake, piloting his canoe under cloudy skies that screened directional information from the Sun (or the stars). Lewis writes that Tevake sat cross-legged, and nearly naked, on the bottom of his canoe, sensing the shape of an ocean swell using his testicles . . .

> He kept course by keeping a particular swell from the east-north-east dead astern, a swell that was effectively masked for me by the steep breaking waves thrown up by the squalls . . . It may seem incredible that a man could find his way across the open Pacific by means of a slight swell that probably had its origin thousands of miles away . . . He had made a perfect landfall . . . having navigated for between 45 and 48 miles without a single glimpse of the sky. [22]

This is very long-distance navigation, and it's the type that can allow for new journeys, as well as return trips to known locations.

We may not all be true explorers, but all of us store sensory memories of where we have been – and perhaps none more so than those that relate to food.

Where did you eat the best burger of your life? Or taste the sweetest strawberry? Or try really hard to savour and not gobble the most delicious cake? You can probably remember not only what the restaurant or food stall looked like and smelled like, but the sounds around it – because, depending on the environment, sounds can also provide valuable texture for our cognitive cartography – and where it was. For our ancestors, memories for the location of wonderful food sources would have enabled much more than happy reminiscing. They would have made the difference between survival and starvation.

Since a range of senses matter for our ability to navigate, individual differences in how good we are at them will affect how well we can find our way around.

There's good evidence that people with vestibular problems can really suffer. 'I think people with vestibular damage rely very heavily on vision and optic flow,' says Kate Jeffery. 'If they close their eyes they can become very disoriented, not just in terms of remaining upright but which way they are facing.'

But even minor vestibular irregularities can cause problems.

Given enough light, we're not too bad at walking a straight course through an unfamiliar environment. But what happens when it gets dark?

If you believe the horror movie *The Blair Witch Project*, we start walking in circles. It's a popular belief, but is it right? In 2007, Jan Souman, a psychologist at the Max Planck Institute for Biological Cybernetics in Tübingen, Germany, got a call from a German science TV show called *Kopfball*, asking exactly this question. Souman had to confess that he didn't know the answer – the necessary studies hadn't been done – but he was intrigued enough to launch his own.

Initial experiments were conducted on blindfolded volunteers who were asked to walk in a straight line across a field. The results revealed that they did indeed start circling, some walking in circles just twenty metres in diameter (while still believing they were walking straight ahead). Sometimes they'd veer off to the left, sometimes to the right. This reflected, the researchers concluded, their increasing uncertainty about where 'straight ahead' was.

For subsequent trials in the Bienwald forest in Germany and the Sahara desert in Tunisia, GPS-tracked volunteers weren't blindfolded. Souman and his team allowed them to walk for hours. And they observed that when the Sun or the Moon was out, these people didn't have much trouble walking in a straight line. But as soon as clouds got in the way, they started circling – and they didn't even notice. They were clearly using the location of the Sun and Moon, even if this wasn't a conscious process.[23]

In a related study, a team of French researchers first tested their volunteers' vestibular functioning. They got them to stand on a force plate, which registers how evenly someone's weight is distributed through their feet. If their posture was perfect, this distribution would have been even. The volunteers also had to position a bar to be vertical, so the researchers could check their subjective sense of 'vertical'. Then they were sent off, blindfolded, to try to walk in a straight line through a vast, empty – and therefore relatively safe – exhibition space in Bordeaux.

The result? A lot of people veered off course. But the greater the imbalance in their initial weight distribution, the more they deviated from a straight line, and the poorer their subjective sense of a straight vertical line. The researchers concluded that very slight irregularities in the vestibular system can skew someone's idea of 'straight ahead' enough to make them walk in circles.[24] Slight imperfections in the physical structure of the vestibular system may, then, help to account for the poor navigational performance of Montello's 'baggers'.

There were no sex differences in this study. Men and women were equally likely to wander off course. In fact, in studies of navigation, men and women will often do equally well – but when a discrepancy is found, it's in favour of the men.[25] Why?

It might be because men seem to prefer mental mapping to route-based navigation. At least, this preference has been observed in studies run by Hegarty and also Sarah Creem-Regehr at the University of Utah, who has found that male subjects tend to take more shortcuts than the women, while women tend to favour familiar routes.[26] Creem-Regehr suggests that this might be because, for our ancestors, taking a shortcut would have been more risky for a woman than a man. If a woman stumbled across a predator's den, for example, she'd have been more vulnerable.

This sounds plausible. But anthropologist Elizabeth Cashdan, also at the University of Utah, wanted to explore what actually happens in traditional small-scale societies. She and her colleagues have studied the Tsimane, who as we already know from the chapter on hearing live in the Bolivian rainforest, and the Twe, who live in open savannah in Namibia. What data largely collected by Helen Davis, now at Harvard University, revealed is that Twe and Tsimane boys and girls are equally good at tests of navigational ability – like pointing from one location towards a concealed other. But though a sex difference in this ability emerges in Twe adults, this is not the case for the Tsimane.

Cashdan has a theory to explain this. Tsimane, who live in dense jungle, tend not to roam far. Men and women have pretty similar ranges. But Twe men roam much further than the women. They travel a long way to visit girlfriends, for example (the society is not monogamous). As a result, they get more experience in navigating than the women. The influence of environmental factors, rather than biological sex, seems to explain the superior performance of the men.[27]

Could this also be true in the West? Even today, men drive more than women, and so get more experience at navigating, says Kate Jeffery. This means that their ability to use and integrate information from the relevant senses is tested more often, and trained.

It should be noted, though, that men also tend to spend more time playing computer games than women. And this could be important for interpreting some research in this area. Men might do better at VR-based navigational tests because they have more experience at operating in a virtual world. (Researchers try to account for this type of thing in their studies, but it isn't always straightforward.) Or that experience could be more meaningful. For instance if you're playing a quest-type game, which requires you to navigate a virtual world and which gives you compass cues as to where you are, this might possibly bring some real-world benefits.

Today, though, how many decent Western navigators could accomplish the types of long-distance treks that Twe men manage (without technology)? Or come even the tiniest bit close to sailing far out of sight of land and relying on touch on their genitals to keep them on track?

As we already know, modern life is dulling many of our senses. But there's also another danger. And the worst bit is that many of

us (myself included) have embraced it, believing it to be our salva-
tion. In fact, it could be more like a body-builder's muscle-aid:
improving things on the surface, but less akin to a wonder-drug
than to a poison.

When we evolved, there were, of course, no map apps to which to
delegate navigation. And this kind of technology harms a person's
ability to form a mental map of an environment, Dan Montello argues.
He doesn't use them himself. 'I am fairly confident that their regular
use impairs a person's ability to wayfind on their own,' he tells me.

In 2019, experimental support for this idea was published by a
team led by psychologist Hugo Spiers at UCL. Spiers and his
colleagues were exploring which regions of the brain were active
when students navigated a simulation of their own campus or the
campus of another, much less familiar university, which they'd spent
time getting to know just a few days before. Some of the students
had directions overlaid on the route before them, so that the team
could also investigate the effects of sat nav.

For the students without the sat-nav-type information, the team
noted that the hippocampus was involved in tracking their virtual
journey to a newly-learned location in the unfamiliar campus.
However, another region, called the retrosplenial cortex, handled
this task as they navigated their own, familiar campus. This work
was important, in showing that two different parts of the brain guide
navigation, depending on whether you're in a well-known or a
relatively new environment. But the team also noted that for both
groups, when the sat-nav direction instructions were given, their
hippocampus/retrosplenial cortex gave up their tracking tasks.[28] 'We
wondered whether navigating a very familiar place would be similar
to using a sat nav, seeing as you don't need to think as much about
where you're going in a familiar place,' commented Spiers. 'However,
the results show this isn't the case; the brain is more engaged in
processing the space when you are using your memory.'[29]

By blunting our natural direction-tracking abilities, sat nav and
phone map apps are, it seems, making us more stupid when it comes
to navigation. It's a great example of 'technological infantalism',
Montello says, adding, 'If you want your children to be able to find

their own ways without nav technology, you will have to let them practise finding their way without it.'

In the modern world, that isn't necessarily easy. But if you are bad at finding your way about and struggle in the context of a modern city, someone just like you would probably have done adequately well in ancestral environments, Jeffery points out. (The question I actually put to her was along the lines of: how on Earth can I, and others like me, have such a bad sense of direction, when it's so important for survival?) Without the clutter of buildings, there would have been ample, large-scale sensory information, providing the types of orientation cues that you just don't get around her office, in central London – mountain ranges, which would always have been visible in daylight; flocks of birds coming from what you knew to be a hidden watering hole; a clear view of the Sun; unbroken shadows. Without looming walls, winding streets or underground journeys, disorientation would have been much less likely.

But given the way many of us live, if you do have a poor sense of direction, what can you do to improve it – aside from ditching map apps?

When you're outside, try to be aware of natural cues. 'If I come out of a train station and I can't be bothered to check my compass, I will look to the Sun,' Jeffery says. And in theory, it should be possible to train people to use shadows, to work out how they are oriented in relation to the points of the compass, adds Hegarty.

Both she and Jeffery recommend simply paying more attention to your environment. As you're walking or driving, make an effort to consciously register landmarks, like a church or a corner shop, and note when you make a turn. You could also try regularly looking behind you, a technique used by some animals, such as digger wasps, which make their nests in bare ground, and form colonies that can include fifty to a hundred separate nests.

There are also some technological training tools that just might help.

Anyone walking through the leafy campus of Mount Holyoke College, in Massachusetts, in 2009 might have caught sight of a faculty member wearing a very odd hat. Sue Barry, the professor of neurobiology whom we met in Chapter 1, had received it as a gift from her husband, Dan, for Mother's Day. Broad-rimmed and striped in shades of brown, pink and yellow, it was, in her own words, 'ridiculous-looking'. She wore it not for sentimental reasons but because thanks to a vibrating device tucked up inside the crown, she was finally learning to overcome something she'd struggled with her entire life. 'My sense of direction ten years ago was pathetic,' Barry confesses. 'I could walk from one place to another by following a certain path and landmarks that I knew well. But I never had any idea in my mind about how they would connect. Do you know *Star Trek*? It was like I was beamed to different places. And it could be very embarrassing. If I had a friend visiting and we'd want to go some place, I'd be driving the car, and I'd have no idea how we were going to get to the museum. I'd have to admit, yes, I've lived in this town for ten years, and it's a relatively small town . . . but I have *no idea*.'

What Dan Barry had created as a gift was a north-sensing hat. He combined an electronic compass, a microprocessor and a motor, which vibrated whenever the front of the hat was facing north. Sue could either hold the motor in her hand, or tuck it up into her hat. 'So, I'd be walking along and when I turned north all of a sudden, I had this vibration. It was a signal coupled with my action – turning to the north – so it made me more aware of, "Oh, I am changing my direction relative to the North Pole." And so, of course, when I was wearing the hat, I was thinking about my sense of direction, because that's why I had the hat on. But then I started to think, "Oh, this landmark is located to the north of that landmark, and this one is located to the south or east or west of some other." I started trying to create a kind of map in my head. I got to the point – I wore the hat often enough – that when I was walking around the college campus, from one building to the next, if I was moving in a northerly direction, I would feel the buzzing even if I didn't have the hat on.'

Her husband's sense of direction is, she says, 'exquisite'. But she soon realised that he, like friends and neighbours with similarly outstanding navigational abilities, were, either consciously or unconsciously, using strategies to work out where they were in relation to the points of the compass. She tested them, asking them while they were at her house to point to the north – and they could all do it. 'I said "How did you know?" One knew where the major north/south road was relative to where we were. One pointed to the sky and said, that's the north star right there – so that's north. Another, a gardener, said, "Well I always get more moss on my rhododendrons on the north side." There are all sorts of ways of doing it. But you have to pay attention. That is key for me. And I think that people who pay attention when they're very young develop habits that they use later on without even knowing that they do.'

For Sue Barry, her use of the hat (and then, later, a phone app that her husband also created for her), along with a new-found appreciation of how to use the location of the Sun in the sky and local landmarks, trained her sense of direction. It's now sufficiently good for her no longer to rely on an artificial crutch. 'I wouldn't say I have a great sense of direction . . . For other people I know, they just "feel" it. For me, it's still a lot of work. But I do have skills now.'

Some researchers are also exploring other ways to make a compass sense feel as natural as possible, using not vibrations, which we don't typically relate to direction, but sounds, which we do perceive as coming from a particular point in space.[30]

For me, looking to the future, my biggest take away from Sue Barry's story, and from the research in the field, is to *pay more attention*. I tend to walk and drive around thinking about all kinds of things that have nothing whatsoever to do with where I am. Now, I at least try not just to notice but to *fix on* sights, sounds, smells, turnings . . . I also try to remember that navigation is a wonderful example of how our brains process multi-sensory data in amazing ways – and while for some of us, this particular process seems effortless, for others, it takes (potentially a lot) more focus. But, unlike long arms, say, it *is* amenable to change.

And, in fact, as a woman, paying more attention to the smells and sounds of my world might help me to get better, faster than a man with the same resolution. The stereotype of men having a better sense of direction than women does have some limited, inconsistent support from academic research. However, as we'll discover in the next chapter, when it comes to sex differences in individual senses, it's almost invariably women who come out on top.

13

The Sex Gap

How men and women sense the world differently

Living, as we do, in our own perceptual bubble, it can be hard to appreciate just how kaleidoscopically different other people's bubbles can be. But though there can be vast differences between individuals, as a group, women sense things differently to men.[1]

In practically every type of sensing that's been tested, women are more sensitive. They – we – are better at distinguishing between colours, we're more responsive to scents, to tastes, to touch, even (contrary to popular opinion) to pain.

Exactly what underpins this sense gap is debated. But because women tend to be better across the board, it's likely to be at least partly due to the way the female nervous system handles sensory signals, rather than differences in the sense organs themselves.

Smell is one sense that historically, at least, has been perceived as being more 'female'. The first scientific hint that women are actually better at smelling came back in 1899, with pioneering French studies that revealed women could detect fainter odours, were better at identifying scents, and found it easier to discriminate between two similar smells – findings that have since been replicated, and observed in girls and boys, as well as adults.[2]

One of these studies represents by far the biggest mass smell experiment to date. In September 1986, the eleven million subscribers, in more than 140 countries, of *National Geographic* opened up their magazine to find a scratch-'n'-sniff card inside. Each card was imprinted with six fragrances: androsterone (a metabolite of testosterone), isoamyl acetate (which has a fruity odour), galaxolide (a synthetic musk), eugenol (a compound found in cinnamon, bay leaves and cloves), mercaptan (rotting cabbages/smelly socks) and rose.

No fewer than 1.42 million readers responded with feedback on which of the six odours they could smell, along with intensity and pleasantness ratings, and some biographical information. About 1,700 even sent back letters about their sense of smell.

Some were amusing. For example:

I thought you would appreciate knowing that after taking the Survey, I presented the six panels to my Golden Retriever. He showed no interest in any of the scents except #5, the one I found most unpleasant. What is it?'

Some were heavy with emotion, such as:

I'm now 85. After my husband died, I missed him so much that I would go into his closet and hug his suits because they smelled of his own body odor, slight cigarette smell, and 'Old Spice.' I'd stand there hugging his clothes, making believe, close my eyes and cry.

Others alluded to female smell superiority. One man wrote that his wife was so sensitive, she should be employed in smell testing. She is an 'expert', he wrote, adding, 'She can smell beer over the telephone.'[3]

The study had been designed in collaboration with two smell researchers at Monell. Among their findings were that women were indeed more sensitive to the scents. Much more recent work has confirmed that women are better at smelling other people, as well as household odours. Given the whiff of someone's underarm odour, they're better at determining whether it came from a man or a woman, and at matching these scents to actual people. When body odours are even 'masked' using different smelly chemicals, this tends to render female odours undetectable to men, whereas it has little impact on a woman's ability to sniff male odours.[4]

For many women, though, their sense of smell is far less constant than it is for men. The stage of the menstrual cycle, pregnancy, menopause . . . all are associated with changes in smell and also taste.

There's evidence that a girl's or a woman's sense of smell becomes even more sensitive in the fertile, 'luteal' period of her menstrual cycle.[5] Around this time, she'll also tend to perceive androstenone (the compound found especially in male body odour, mentioned in Chapter 3) as being less unpleasant. What's more, another 'male' body odour, androstadienone, improves women's perceptions of the attractiveness of male faces and voices – but only during this fertile period.

In fact, all kinds of sensory tweaks and peaks have been observed over the course of the menstrual cycle. A team of physiologists at the Himalayan Institute of Medical Science in India is just one to have found changing salt preferences. In their study, women at various stages of their menstrual cycle were given samples of popcorn that had been sprayed with salt solutions of varying strengths, and asked to rate how much they liked them. Those who were menstruating were keenest on the unsprayed, unsalted batch. However, for women in the fertile phase, it was a case of the saltier the better.[6] During this fertile period, there are also hints that at least some women find meat less palatable. Sour-taste sensitivity seems to decrease, too, in theory perhaps making women want to eat more in the way of sour-tasting foods.[7]

Why might these taste changes happen? No one knows for sure. But of all the food groups, meat is the most likely to carry an infection, a threat to a potential pregnancy. As for the sour finding, sour fruits, such as citrus fruits, also tend to be high in sugar, an easy source of energy – and there's some evidence that around ovulation, women tend to develop a sweeter tooth, too. Salt, meanwhile, is needed for maintaining water in the body, and women have to retain extra water during pregnancy.

When it comes to obvious cravings during the menstrual cycle, though, most women probably wouldn't point to the fertile phase, but to the pre-menstrual period – and chocolate. There's some debate about what drives these cravings – is it biology or culture, or both, or neither?[8] Only more work will tell.

But there is a state of being often associated with extreme, and even bizarre, food desires, as well as other shifts in the senses, and that has been better studied, all the way down to the receptor level, at least in animals. That's pregnancy, of course.

*

My second pregnancy was a[n] olfactory nightmare. The dog stank to high heaven, my firstborn was a diaper-wearing terror of wafting fumes, and I actually woke my husband up from a sound sleep to make him go brush his teeth in the middle of the night. Really.

I've always had a sensitive nose and it was magnified by my pregnancy as well. Horrible. I sometimes find being out in public overwhelming with all the perfumes and body odors and whatnot.

These reports from pregnant women were collected by psychologist Leslie Cameron at Carthage College, USA, for a review of studies into pregnancy and smelling.[9] As Cameron notes, the idea that smell is heightened in pregnancy goes back at least to 1895, to a paper by a Dutch scientist called Hendrik Zwaardemaker. According to research in the twenty-first century, about two-thirds of pregnant women report feeling that their sense of smell has become keener. In particular, they talk about a stronger reaction to smells during the first trimester. These observations have led to a theory that pregnancy sharpens a woman's sense of smell to protect the foetus – as this helps her to sniff out rotten or otherwise toxic food.

It all sounds perfectly neat. However, the data are messy.

According to some studies, pregnant women report that their sense of smell has got *worse*, while others have found no effect. Part of the reason for inconsistencies in the results could be that when pregnant women do report that they are more sensitive to smells, they don't necessarily mean it in the way that smell researchers do.

In lab investigations of how much of a smelly substance has to be there for someone to be able to perceive it, generally pregnant women don't do any better than other women. Technically, their smell sensitivity doesn't seem to be enhanced. Neither, according to another set of studies, is their ability to identify scents – to report that one whiff is cinnamon, for example, while another is cloves. However, when you ask women whether certain odours *seem* more, or less pleasant, this is when clear differences emerge. During pregnancy, women tend to find a variety of odours less agreeable, or more unpleasant – or even, in fewer cases, more appealing.

Among the smells often rated during pregnancy as being less pleasant, or even downright repugnant, are those given off by meat, fish, eggs, garbage, burnt food, cigarettes, human bodies, perfumes and colognes. The 'more pleasant' list is much shorter, and includes pickles, fruits and spices.

Women in the first trimester of pregnancy do tend to have a stronger general disgust response. As we know, disgust evolved to motivate us to avoid potentially poisonous foods. Stronger feelings of disgust could then make for stronger aversions to foods that carry a high infection risk, such as meat and eggs, in particular. But, if a woman's sense of smell typically doesn't get more sensitive during pregnancy, what might explain a sudden abhorrence of perfumes and colognes?

There is a theory, and it goes like this:

Morning sickness is common in the first trimester, especially. As we are so driven to avoid things that make us sick, when we do vomit, our brains take a comprehensive inventory of the sensory scene and store it away for future reference. Personally, I still have a strong aversion to a brand of tinned chicken soup that I was invariably given when I was ill as a child. I know it wasn't the soup that made me sick. But my unconscious brain is having none of it; a poor detective, it's convinced it has to be the culprit because whenever I was feeling rotten, that particular soup was *always* there.

Now back to pregnancy: a woman who's often feeling nauseous and even vomiting is going to learn quickly to pair a range of sensory signals – and smells and tastes are obvious ones – with a rolling stomach and losing her lunch. It's in our survival interests to *over*-pair, to err on the side of extreme caution, and, if your brain thinks you *may* have consumed something toxic, to get it out (which is why if one kid vomits on the school bus, others will quickly follow suit). In this way, even a scent that a woman didn't consider at all offensive before she was pregnant could become distinctly off-putting. Stronger and more regularly encountered smells – such as friends' perfumes or even the kitchen bin – would be more likely than milder scents to make their way into this classical conditioning equation.

Smell isn't the only sense that pregnant woman feel alters. About nine in ten will say that things taste different, too.[10] Often, this entails feeling more sensitive to bitter-tasting foods, and less sensitive to sweet ones. However, studies that have used actual taste function tests have produced very mixed results. Some find that pregnant women are more responsive to salt; others don't. Some observe less of a sensitivity to sweet compounds; again, others don't.

What could explain this? It could be to do with the huge natural variation between people in taste sensitivities. In some studies, strong differences between individuals may be overwhelming or at least muddying the evidence of group findings.

To understand how pregnancy may affect taste, it's helpful to get down to the level of the taste receptors themselves, and to explore how pregnancy-related hormones might change how they function.

We know that human taste buds contain receptors for various hormones, and that list includes oxytocin. Dubbed the 'cuddle hormone', oxytocin is essential for bonding between a mother and her baby, and also labour and lactation. (It's a release of oxytocin in response to the touch of the infant's mouth on the nipple, or even the sound of a baby crying, that triggers the milk 'let-down' reflex.) Levels of oxytocin increase gradually during pregnancy, and there's some evidence from animal studies that it influences sweet-taste sensitivity, in theory perhaps encouraging the woman to eat more sweet things, and so drive a higher calorie intake.[11]

Another hormone, angiotensin 2, which is important for regulating blood pressure – and so is critical for the health of the mother and foetus – seems to decrease salty-taste sensitivities (at least, this has been shown clearly in mice).[12] In theory, this might encourage a pregnant woman to consume sufficient salt to maintain a rise in blood fluids, and keep her blood pressure stable.

Hormonal changes during pregnancy could conceivably also have an impact on how the brain processes sensory information. But there are no data at this level in people – because no one has looked. It should be noted, though, that sensory processing is not a particularly neglected area in the field of research on the brains of pregnant women. The *entire field* has been neglected. And as far as some researchers in this tiny, nascent community are concerned, the scarcity of scientific research into the impacts of pregnancy on women's brains isn't so much embarrassing as scandalous. For an event of such startlingly transformative magnitude – one that unleashes a tsunami of hormones, with levels of oestrodial, as just one example, soaring to levels hundreds of times higher than normal, not to mention its being something that happens at least *211 million times* in the world every year – it's an area that's desperately under-studied.[13]

The data we do have come from animals, and observational work on their behaviour has found some startling differences. These studies suggest that having offspring sharpens abilities that are important for ensuring their survival, and that this involves the senses. For example, mother rats are better than those that have never been mothers at remembering the location of food in complex mazes. What's more, they are stunningly better at catching prey.

Craig Kinsley, a neuroscientist at the University of Richmond in Virginia, who became passionately interested in the potentially positive impacts of pregnancy on the brain after watching his own wife caring for their newborn baby, while maintaining all her previous tasks, has found that while non-mother rats take, on average, 270 seconds to hunt down a cricket hidden in an enclosure, mothers take just over fifty seconds.[14]

Human mothers generally don't go off to hunt animals to feed to their babies. But there is evidence that, like rat mothers, they tend to become more aggressive when provoked. Starting in late pregnancy, women also get better at registering facial signals of fear, anger and disgust, though their ability to detect surprise and positive emotions does not change. This makes sense, says Laura Glynn at Chapman University, USA, who led this work: 'If you're trying to protect your infant, you want to be able to detect a threat.'

Hints that at least some pregnancy-related changes may persist have also emerged. Certainly, there's some evidence that becoming a mother changes the way a woman's brain responds to hormones for years afterwards.[15] But what this might mean for brain function, in general, and sensing, in particular . . . no one really knows. For Glynn, this lack of knowledge is 'almost a crisis in women's health. How can we *not* know the answers?'

If pregnancy were not challenging enough, there is of course the birth – and the pain that comes with it. There is a popular belief that women are less sensitive to pain, or at least more stoical about it. In scientific circles, this idea was challenged, and it became unacceptable to suggest that there are any sex differences. However, in 2004, a team of pain specialists at the University of Bath reviewed work on sex variations in pain and concluded: 'Until fairly recently it was controversial to suggest that there were any differences between males and females in the perception and experience of pain, but that is no longer the case.'[16] Only, the evidence contradicts the folk belief.

When it comes to extreme temperatures – both hot and cold – women are more responsive, being faster to feel the burn or the chill, and feeling it more strongly. This group difference extends to other potentially damaging impacts, too, and so to pain perceptions. The Bath team based their conclusions on a range of research – from lab studies assessing pain thresholds to field studies in hospitals finding that women tend to report more pain over their lifetimes, in more parts of their body, more often, and for a longer duration.

Certainly, women are more likely than men to report chronic pain – pain that persists for longer than twelve weeks, despite treatment. Damage to nerves can be responsible, and recent work has found that when this kind of damage happens, the immune systems of men and women don't respond in the same way. Remarkably, there are differences in the types of immune cell implicated in the response for each sex. There are also strong hints that this difference between the 'male' and 'female' pain mechanism, which seems usually to be driven by higher levels of testosterone in most men, at least partly explains why women suffer more from chronic pain.[17]

There's also a fascinating link between chronic pain and pregnancy. Many women with chronic pain who become pregnant report an easing of their symptoms. And in pregnant mice, at least, researchers have observed that early on, there's a switch from the 'female' pain mechanism to the 'male' one. Strikingly, the researchers report that by late pregnancy, female mice show no evidence of chronic pain whatsoever. They suspect that in this case, rising levels of the hormones oestrogen and progesterone are responsible for the 'male' shift.

The research still has a long way to go, however, and the historic lack of scientific studies focusing on women and their senses has everyday implications even now. We already know from Chapter 10 that the typical office isn't necessarily a comfortable environment for women, because of temperature guidelines alone. But if the sound of an office printer or a coffee machine seems 'too loud', and there is disagreement with male colleagues about this as well, this is another example of a documented sex difference in sensing. As a group, women are more sensitive to sounds.[18]

It's not all bad news, though. Gary Lewin and his group have found that women are about 10 per cent more sensitive to touch. It's the same for young adults, too. If this difference exists in children – and there are no data on this yet – it might possibly account for some of the differences in key markers of progress in boys and girls. If girls do have a better sense of touch, it could be a little easier for them to learn to wield a pencil effectively, and so develop better, neater handwriting than boys. Then there's talking. To form clear speech sounds, you need to sense precisely which part of your oral cavity is being touched by your tongue. Various reasons have been put forward for why girls typically learn to talk earlier than boys. Better touch sensitivity in the mouth could easily be important.

Young girls are also typically quicker to learn to name colours correctly and consistently. What's more, as women, they're better at naming colours from memory than males – and also seem to see colours a little differently.[19] US research suggests that for men and women to perceive the same colour, the wavelength has to be a little longer for men. This could mean, for example, that 'green' grass looks a little yellower to a man.[20]

If these results seem intriguing rather than satisfyingly fleshed-out, well, that's the current state of affairs. We are an awfully long way away from fully characterising and understanding sex differences in sensory processing and sensory perceptions. Whatever the underlying mechanisms, though, as groups, women and men are different, and women do tend to be more sensitive. In some situations, greater sensitivity can be seen as a 'good thing'. Women are much more likely to find themselves on professional smelling and taste panels, for example, than men.[21] In others – in pain, for example – not so much.

But when it comes to how some groups sense differently to others, and what this means for their lives, there are sensory tribes that cut across the sexes and for whom the impacts are immense. In the next chapter, I look at how our senses are critical for our ability to feel not only emotion, but empathy. It's arguably in this sphere that who we are is most in thrall to how we sense.

14

Sensing Emotion

How our senses make our emotions

Your phone's ringing. You answer it, and it's your boss, with terrible news: you're losing your job. Immediately, your pulse speeds, you feel sick, your breathing quickens, your palms sweat . . . If you had a stress expert to hand, they'd tell you to use the stopwatch function on your phone to regulate your breathing, to slow your in-breaths and especially your out-breaths. Do this, and within a few minutes you will have calmed down. Many of us know that this kind of technique eases acute anxiety. But think for a moment about what it reveals.

The *thought* that stressed you out has not changed. You're still going to lose your job. All that's altered are the interoceptive messages that receptors in your lungs are sending to your brain. Your *body* is now signalling that you aren't so anxious any more – and your brain is believing it.

Aristotle thought that emotions originate in body states: 'The brain is not responsible for any of the sensations . . . The motions of pleasure and pain and generally all sensation plainly have their source in the heart.' It's an idea that we express explicitly in everyday language. When you truly love someone, it's with 'all your heart'. For an apology to mean anything, it has to be 'heartfelt'. A 'hearty' welcome is a genuine one. Being dumped can leave you 'heart-broken'.

But though the idea that the heart is the seat of emotion persisted in language, in philosophy, it did not. In the seventeenth century, Descartes sundered the mind from the body, and this mistaken separation held sway for centuries. Then in 1872, Charles Darwin published *The Expression of the Emotions in Man and Animals*. In this book, he argued that many different species (a list that included humans) express emotions in remarkably similar ways. Before long, Carl Lange, a Danish physician, and also William James were independently arguing that bodily signals underpin our emotions. In an 1884 paper titled 'What is an Emotion?' James maintained that, should something 'exciting' happen, 'the bodily changes follow directly the perception of the exciting fact . . . our feelings of the same changes as they occur is the emotion.'[1]

In *The Principles of Psychology* (1890), James reiterated his idea that the perception of these bodily changes 'IS the emotion' (his capitals). He goes on:

> Common sense says, we lose our fortune, are sorry and weep; we meet a bear, are frightened and run; we are insulted by a rival, are angry and strike . . . The more rational statement is that we feel sorry *because* we cry, afraid *because* we tremble. [my italics][2]

At first read, this might seem counter-intuitive. Surely you have to recognise a threat or an exciting or scary event before your body can respond? Of course, that's true − but that recognition doesn't have to be conscious.

The thalamus, which forwards on incoming sensory data, has a hotline to the amygdala. And the basolateral region of your amygdala is also a first-line responder to signals from sensory processing parts of the cortex. It automatically registers sensory signals indicating that something is of biological relevance to you. This can certainly be a threatening something (like a bear), or, alternatively, something that enhances your survival chances, such as the visual signals of a slice of chocolate cake, or a friend approaching you in a café, because maintaining social relationships is important.[3]

To take James's bear example, if the amygdala does identify a clear and present danger, it despatches messages that gear you up for mounting a response − for fight or flight. As we know, this entails activation of the hypothalamic-pituitary-adrenal axis (HPA axis), causing your adrenal glands to pump out adrenaline, which, among other things, raises your heart rate and relaxes the smooth muscle in your lungs, allowing you to absorb more oxygen.[4] Mechanoreceptors in the heart muscle, lungs and in your blood vessels register these changes, and their signals are received by the insular cortex, a region known to process bodily sensory signals *and* to be important for emotion. Only once you've registered these inner bodily messages, the theory goes, do you *feel* afraid.

If you're not convinced, think about what fear *is*, without the rapid breathing, shaking and pounding heart. For that matter, what's left of anger, James writes, if there is no 'ebullition of it in the chest, no flushing of the face, no dilatation of the nostrils, no clenching of the teeth, no impulse to vigorous action, but in their stead limp muscles, calm breathing, and a placid face? . . . The rage is as completely evaporated as the sensation of its so-called manifestations.'

These arguments didn't sit well with everyone. For example, Walter Cannon didn't believe that sensations from the body *caused* emotions.[5] However, James and Lange's ideas have held, though the theory of emotion rooted in their ideas has been adapted, to allow for your brain's predictions about what your bodily states will be in a given scenario to influence how you feel.[6]

Some researchers have been sceptical about the idea that our varied emotions have corresponding, unique patterns of bodily sensations. But there has been a recent softening in this opposition, from some, at least.

In 2018, a group led by Lauri Nummenmaa, at the University of Turku, Finland, asked more than 1,000 people where, in their bodies, they *felt* more than 100 emotions and other states. The list of states was very broad. It included, for example, gratitude, fear, love, guilt, social exclusion, drunkenness, despair, pride, reasoning and recollecting. The team then combined these responses to produce a beautiful chart of coloured bodies for different states, with bright yellow spots for the regions that the respondents felt were most affected, moving to red and finally to black for parts that weren't involved at all.[7]

Strikingly, there was a high level of agreement among the participants about what they felt where. Also, almost all of the feeling states had a unique, discernible body map – a 'body sensation fingerprint'. These maps tended to include sensations both from organs and from muscles – so proprioceptive sensations, too. For 'relaxation', for example, the entire body is red, with yellow patches in the shoulders and arms. For anger, the hands are yellow, and there's a yellow blob high in the skull. For happiness, there's a yellow glow in the heart region, and a bright yellow suffusing almost the entire head, with traces of red along the arms. Nummenmaa believes that the work demonstrates that conscious feelings (including, but not restricted to, emotions) stem from bodily feedback – from inner-sensing and also touch and body-mapping, proprioceptive signals.

The people who took part in this Finnish research were not professional experts in emotion by any means. But some of their ideas about where we feel emotions have been supported by lab studies. For example, whether you're angry or afraid, your heart rate rises. However, with anger, there's increased blood flow to the hands (to prepare you for using them – perhaps to punch someone else), but this isn't the case with fear. Also, with fear, blood flow is diverted away from the face, whereas with anger, the face flushes red with blood. And though a faster pulse is common to both states, it's well known that among people who've experienced a heart attack during strong emotion, it's much more likely to have been during anger than fear. Something about the way the heart is regulated in anger has to be different to the way it is in fear, concludes Hugo Critchley at the University of Sussex.

Critchley and Sarah Garfinkel think that we unconsciously recognise patterns of inner bodily signals, and use this information to feed into our perceptions of emotion in the way that we use patterns of taste and odour receptor stimulation to build up perceptions of the flavours of a meal.[8] We are truly exceptional spotters of patterns in the wider world. It would be odd if we didn't learn to use patterns in bodily signals to our advantage, too. (And in fact, the pair's work on the gut feelings of London traders, described in Chapter 12, shows that we do.)

But, you might be thinking, if unconscious processes can handle the detection of objects or events that represent a threat or a boon to our ability to survive and thrive, why do we need to *consciously* feel emotions? The answer is that unconscious recognition is one thing. But emotions grab us. They force us to pay attention, and drive us to use all our resources to come up with the best possible response. For example, anger might drive us to hit someone – but if we can control that impulse, it might alternatively lead us to strive to find non-violent ways to force compliance. And if anger drives you to change something that really needs changing, it's intensely valuable. (Aristotle had a nice line on rage: it was, he wrote, 'noble, just, useful and sweet'.[9]) A feeling of joy at seeing a friend, meanwhile, puts you in the right mental state for further cementing that relationship, which could help you to weather future threats.

In an ideal world, we all accurately read our bodily sensations, immediately apprehend the full context of every encounter, have an ample lexicon of emotion terms to enable fine-scale appreciation of our state, and are in perfect tune with how we're feeling.

But imagine someone who only ever drinks decaf coffee arriving to take part in a business negotiation and accidentally pouring herself a caffeinated cup. (It's easy to mix up the flasks, isn't it? I've certainly done it.) As a result, her heart starts to race, but as she has no idea of her mistake she concludes that she's feeling anxious, and decides that it's because she's under-prepared. This threatening thought stokes genuine anxiety, which in turn makes her performance in the meeting much less confident and persuasive than it would have been without the self-slipped, emotion-warping Mickey Finn.

Such mix-ups with coffee are unlikely to happen often. But it's relatively easy to imagine how – especially if the sensations are faint, or there's significant overlap between the body sensation patterns associated with two emotions, and the context is complex – we may plump for the wrong emotion label. In 1974, the psychologists Arthur Aron and Donald Dutton demonstrated that this certainly can happen. Their now classic study quickly became a favourite among psychology lecturers, in part because it allows them to spice up a course on emotion with something very relevant to many students: a recommendation for the perfect first date.

The field experiment was conducted in North Vancouver, on men, aged between eighteen and thirty-five, who were crossing either of two bridges. The Capilano Canyon suspension bridge has very low handrails, a tendency to sway and wobble, and a 230-foot drop to rocks and rapids below. This is, by any estimation, a 'scary' bridge. The other bridge is solid, wide and sturdy and positioned just ten feet above a nearby rivulet. It isn't scary at all.

Research assistants (an 'attractive' male or female interviewer) lay in wait on each of the bridges. When they spotted a man crossing, they approached him, and asked him to complete a questionnaire. Afterwards, the interviewer scribbled their name and number on a corner of the page, tore it off, and proffered it, telling the man to call if he wanted to discuss the study further.

Aron and Dutton found that men who'd met the female (but not the male) interviewer on the scary suspension bridge were much more likely to call her than those who'd encountered her on the solid, sturdy bridge. They concluded that the first group had interpreted at least some of their fear-related physiological arousal at walking over the wobbly bridge as sexual attraction. (The assumption was that they were all heterosexual.) The work indeed suggests that a thriller or a horror film or a day spent on rollercoasters could make a good first date. But it also demonstrates that our body sensation fingerprint-detection system doesn't quite match that of a top detective. Occasionally, we'll ID the wrong guy.[10]

In some situations, bodily signals can even lead us seriously astray, with potentially disastrous consequences.

Fear is associated with intense alertness, and also a tendency, especially in situations of perceptual uncertainty, to err on the side of caution. That rustle in the grass? More likely to be an attacker than the breeze. The object in that Black man's hand? More likely to be a gun than a mobile phone. This shocking finding comes from research by Critchley, Garfinkel and colleagues, published in 2017.[11] Pictures of Black or white individuals holding either a gun or a mobile phone were flashed to people either *on* their heart contraction (when heart beat baroreceptors fire off signals), or between contractions. When the images were shown on the heart beat, the participants were more likely to *perceive* the object as being a gun when it was held by a Black person.

The main message from this work, the researchers say, is that faster, stronger heart beats – the sort triggered by the amygdala when it registers a danger, and which are signalled back to the brain as bodily sensory data – will increase the chance that you'll perceive something innocuous as being threatening; *and,* if you tend to believe that Black people are more likely than white people to carry guns, a gun is what you may well *see.*

The researchers think that such errors of judgement, due to

high heart-rate signals, could help to explain the relatively high number of shootings of unarmed Black versus white people. According to the *Washington Post*'s constantly updated police shootings database, every year for the past five years, US police have shot and killed about 1,000 people, most of whom were armed. A 2017 analysis found, though, that 15 per cent of African Americans versus 6 per cent of white people killed by police in 2015 were unarmed. The more recent data show that rates of US police shootings of unarmed people have fallen significantly since 2015, but Black people, and also Hispanics, are still disproportionately represented on that list.[12]

Some people are of course better than others at regulating their emotions and their actions – at *not* flaring up and not reacting impulsively. This is largely due to stronger restraining connections between the prefrontal cortex and the amygdala. These connections develop during childhood and into early adulthood – and some of us develop stronger cortical control over our emotions than others.

But how strongly we feel bodily signals in the first place is also important. Just as your neighbour's jasmine might smell overwhelmingly strong to you while he barely notices it, you may be more sensitive to bodily signals relating to emotions. This kind of difference really matters. It could even determine how well the pair of you get on. Because where we each sit on the spectrum of inner-sensing ability and awareness profoundly affects – even determines – our emotional lives.

Stephen has been married twice. Two wedding days. Two 'I do's. Yet he has no happy memories from either – or, in fact, from his marriages, or any of his relationships.

Stephen met his first wife on a pre-nursing course when he was just sixteen. Six years later, they were married. Three years after that, they got divorced; she was never really the right one for him, he says. Almost two decades on, in 2009, he met his second wife through a dating site. He threw himself into the relationship and, the following year, with his father and her two adult siblings present, they married at the registrar's office in Sheffield, where they both live – and where Stephen and I are now talking, in a cinema café.

He put on smiles for the wedding photos because he recognised that they were expected but, he says, 'From an inner feeling point of view, anything I do that requires an emotional response feels like a fake. Most of my responses are learned responses. In an environment where everyone is being jolly and happy, it feels like I'm lying. Acting. Which I am . . . So it is a lie.'

Happiness isn't the only emotion that Stephen struggles with. Excitement, shame, disgust, anticipation, even love . . . he doesn't feel these, either. 'I feel something but I'm unable to distinguish in any real way what that feeling is.' The only emotions he is familiar with are fear and anger.

Such profound problems with emotion are sometimes associated with autism, which Stephen does not have, or with psychopathy, which he doesn't have, either. Recently, at the age of fifty-one, he finally learned what he does have: a little-known condition called alexithymia (pronounced alexi-thigh-mia, from the Greek *a* (no), *lexis* (word) and *thymos* (emotions). Surprisingly, given how generally unrecognised it is, studies show that about one in ten adults fall on the alexithymia spectrum. That means we all are likely to know someone who's on it.

Despite the name, the real problem for people with alexithymia isn't so much that they have no words for their emotions, but that they lack the emotions themselves. Still, not everyone with the condition has the same experiences. Some have gaps and distortions in the typical emotional repertoire, or can come across as emotionally 'flat'. Some realise they're feeling an emotion, but can't tell which, while others confuse signals associated with one state for something else – perhaps interpreting butterflies in the stomach as hunger pangs.

The term itself dates from a book published in 1972, and has its origins in Freudian psychodynamic literature.[13] Freudian ideas are now out of favour with most academic psychologists, as Geoff Bird, professor of psychology at the University of Oxford, explains: 'Not to disrespect those traditions, but in the cognitive, neuro, experimental field, not so many people are really very interested in anything associated with Freud any more.'

But when Bird read about people with alexithymia, he found the descriptions intriguing. 'Actually, it's really quite amazing. For most people, at a low level of emotion, you might be a bit unsure about exactly what you're feeling, but if you have a strong emotion, you usually *know* what it is.' And yet somehow, here were people who simply did not.

Bird has since run a series of studies exploring alexithymia. He has found, for example, that people with the condition have no trouble recognising faces, or distinguishing pictures of people smiling and frowning. 'But for a few of our really alexithymic people, while they can tell a smile and a frown apart, they have no idea what they *are*. That is really quite strange.' For Stephen, while he can certainly recognise a smile, and respond in kind, there's a lag. His smile as I approached the café table at which he was sitting didn't come automatically. I could see that he had to consciously note mine, and that he chose to echo it.

Many of the people with the condition whom Bird has met talk about being told by other people that they're different, though some do recognise it in themselves early on. 'I guess it's a bit like not being able to see colour,' Bird says, 'and everybody's always banging on about how red this is or how blue, and you come to realise there's an aspect of human experience that you're just not participating in.'

As well as better characterising alexithymia, Bird and his colleagues have also dug into what explains it. In situations that Stephen recognises as being in theory highly emotional – such as telling someone, 'I love you' – he does perceive some changes inside his body. 'I feel my heart race and this rush of adrenaline, but to me that feeling is always scary. I don't know how to react. It makes me want to either run away or react verbally aggressively.' Fear and anger – and confusion – he understands. 'Everything else just feels all the same . . . it's this feeling of, "Errrr, I'm not quite comfortable with this – it's not quite right."'

For Rebecca Brewer (a former student of Bird's, now at Royal Holloway, University of London) this makes sense. 'With alexithymia, people often know that they are experiencing an emotion but don't know which emotion it is. This means they could still experience depression, possibly because they struggle to differentiate between different negative emotions, and struggle to identify positive emotions.

Similarly with anxiety, it might be that someone experiences an emotional response associated with a fast heart beat – which might be excitement – but they don't know how to interpret that, and they could panic about what's happening in their body.'

What Bird, Brewer and others have found in people with alexithymia is a reduced ability – sometimes a complete inability – to produce, detect or interpret these bodily changes.[14] People with this condition have IQs within the normal range and can understand as well as anyone else when they're seeing a spider, rather than an attractive potential partner, or a coffee cup. But either they just aren't undergoing the bodily changes that it seems are needed for the experience of an emotion, or – as appears to be the case for Stephen – they aren't reading these interoceptive signals properly. In 2016, Bird and Brewer, along with Richard Cook at City University in London, published a research paper that characterised alexithymia as a 'general deficit of interoception'.[15]

People at the other end of the spectrum – those who do well at the inner-sensing tests described in Chapter 9 – tend to feel emotions more intensely, and experience more nuanced emotions. In fact, such people are better able to recognise not only their own but others' emotions, a crucial first step in being empathetic. Conversely, people like Stephen struggle not just with their own emotions, but with empathy, too.

That doesn't mean he doesn't care about other people. He has no trouble understanding that an employee whose close relative has just died, say, may be struggling and need time off work. But for someone who isn't alexithymic, when we rejoice with a friend over a new baby or weep at a news photograph of a bereaved child in a war-torn city we experience these emotions not because we're thinking about how they must be feeling, but because we are automatically feeling 'with' them. How empathy happens, though, requires the telling of a bigger story of people-sensing.

Elaine Hatfield is a professor of psychology at the University of Hawaii. In the late 1980s, she was working as a therapist, seeing clients together with a colleague, Richard Rapson. She and Rapson started to talk about how easily they both caught the 'rhythms' of their clients' feelings from moment to moment – and, as a consequence, just how much their moods could change from client to client, and hour to hour.

In a subsequent book, Hatfield writes about how, in the company of a depressed patient, she'd feel a 'dead, sleepy feeling' come on. 'I am so prone to the deadening effects of the depressed,' she writes, 'that it's hard to keep even a minimal conversation going with them; I keep finding myself sinking off into sleep.'[16]

These kinds of observations led the pair, in collaboration with psychologist John Cacioppo (who became famous for pioneering research on loneliness) to examine the foundations of this: how one howling dog will set off another, or how if a baby starts crying in a ward of newborns, the rest will quickly follow. They labelled this automatic, unconscious process 'emotional contagion'.

Emotional contagion is thought to be the evolutionarily primitive basis of empathy, and it's clearly important for survival. Hear a scream, and you're immediately put on high alert, ready to fight or run. Observe a new colleague coming towards you with a broad smile, and the opposite happens: you recognise that here's a potential ally and friend, and you start to feel warmly towards them, too.

Though vision and hearing are important for sensing the emotional state of others, so too is smell. Pam Dalton at Monell has investigated the impact of a person's own emotional state on their body odour. As we know from Chapter 4, our breath, our urine and even our blood contain molecules that smell, but the main source of body odour is the underarms. Dalton and her colleagues collected underarm samples from people who'd been made to feel stressed in the lab. Another group smelled these odours while they watched videos of women doing something that might potentially be stressful – like getting kids ready for school while trying to cook breakfast – but where the women's faces, movements and posture didn't suggest that they were stressed. ('We combed through hundreds – if not more – videos to try to find a set that worked!' Dalton remembers.)

When the people watching the videos smelled the 'stressed' body odours they rated the women in the videos as being more stressed than if they smelled a mild neutral fragrance or underarm samples collected from volunteers who'd been exercising. With the 'stressed' odours, male viewers (but not female ones) also rated the women as being less trustworthy, less competent and less confident. And yet the viewers didn't rate any of the three odours as being more or less pleasant, or even as being very different from the others. The team concluded that subconscious signalling was going on.[17]

That study, published in 2013, helped to trigger all kinds of other experiments in the field. In 2018, a team at the Max Planck Institute for Chemistry in Mainz published details of an 'objective' method for deciding which age rating to give to a film.[18] The team measured the air composition of compounds given off by human bodies during 135 screenings of eleven different films. In total, more than 13,000 people took part. The researchers found that air levels of one particular chemical, called isoprene, reliably correlated with the age ratings given to the films. 'Isoprene appears to be a good indicator for emotional tension within a group,' commented team leader, Jonathan Williams. And it was clear why this might be the case. Isoprene is stored in muscle tissue and released when we move – while squirming in a cinema seat, for example, or tensing our muscles in fear or excitement.

That same year, an Italian team published work revealing potentially grave implications of smelling someone else's fear. They reported that dental students who performed treatments on medical mannequins dressed in T-shirts that had previously been worn by students during a stressful exam did worse than students working on mannequins wearing T-shirts that had been worn during a chilled-out lecture. (Worse was bad; 'healthy' teeth were damaged.)[19] The dental students had apparently 'caught' the stress experienced earlier by their fellow students, and this blighted their performance. Precisely which compounds we use, day to day, to smell fear, or other emotions, in others is still being explored. But it's worth bearing in mind that whatever these chemicals are, if you're producing them, you're inhaling them, too.

Let's say you're about to sit an important exam, or give a presentation to a large group of people. If you're like me, you might feel nervous just at the thought. But there could be one way to help keep a cool head: deodorant. Because she doesn't feel she has much body odour, Dalton doesn't usually wear deodorant. But if she knows she'll be going into a stressful situation, she puts some on. That's because she wants to protect herself from potentially psychologically-damaging scent signals that her own body may produce. Dalton thinks we could all benefit from a better understanding of how smells can affect us: 'If we're not aware we're being influenced, we can't guard against it,' she explains.

So, we use vision, smell, hearing, and potentially also touch, to glean information about other people's emotional states. But how does that allow us to *share* in those emotions – to beam with genuine joy at a friend's good news, or to cry with them at their loss? Neuroscientists and (married) research partners Christian Keysers and Valeria Gazzola have a persuasive theory about this.

About ten kilometres from downtown Amsterdam is the Netherlands Institute for Neuroscience, and the Social Brain Lab, which the pair jointly head. In the Lab's main space are around twenty desks, an impressive coffee-making machine and an antique, upright piano, with a book of Schubert open on the stand. Gazzola is the piano-player. 'I'm more of a Chopin and Beethoven person,' she says, smiling. 'Chris's dad found the Schubert in a second-hand market and gave it to me. I haven't tried it yet.'

But if someone were to walk in now and play the Schubert, Gazzola wouldn't only hear the music, but, as she watched the pianist's fingers on the keys, she'd process the motions in her premotor cortex. This part of the brain prepares your muscles for movement. And there's evidence to suggest that it's part of a 'mirror system' – a system that gives you a kind of first-hand understanding of what someone else is doing physically. Extending this idea, Keysers and Gazzola think that a mirror system allows you to simulate how someone else is feeling emotionally – and that this plays a key role in empathy.

This research started with monkeys and raisins and a talk given in 1999 while Keysers was a PhD student at the University of St Andrews in Scotland. Neurophysiologist Vittorio Gallese, from the University of Parma, came over as a visiting speaker. He spoke about how, in 1990, he and his team had identified a 'mirror neuron' in the premotor cortex of macaque monkeys. When a monkey picked up a raisin from a tray, this neuron fired in its brain. When the monkey watched Gallese pick up a raisin from the tray, it also fired. Whether the monkey was doing the action itself, or watching another monkey, or a person, do it, the same neuron responded.[20] During his talk, Gallese put forward the idea that perhaps mirror neurons might give us an intuitive understanding of the physical actions of other people.

Keysers was fascinated by the idea. Two weeks after handing in his PhD thesis, he headed straight to Parma, to work with the group. In further studies with the macaques, the team found that about 10 per cent of their premotor neurons are mirror neurons, and they respond to slightly different things. Take eating peanuts as an example. Some neurons will become active when a macaque, or another monkey or human, breaks the shell, others when the peanut is picked out of the shell and yet others when it is raised to the mouth. The pattern of their activity acts as a record of the sequence of individual actions that culminate in a nut being consumed – a kind of muscle-control version of a computer programme, which allows the monkey to copy these actions, and so learn by observation.[21]

To date, mirror neurons have been found in seven different regions of the macaque brain: in two regions of the premotor cortex, in regions that deal with eye movements and also in the somatosensory cortex, which of course processes touch and proprioceptive signals.

Vicariously experiencing the actions of another speeds a monkey's ability to master a novel task, like how to get peanuts out of a shell. In theory, this would also be useful for us, of course. An implicit understanding of what someone else's body is doing could also aid activities in which people have to work together – as in hunting, playing a game of football, or dancing.

The old adage still stands: if you want to get better at a task, practice makes perfect. But there's evidence that as we practise these kinds of tasks, we do get better at predicting not only the results of our own bodies' actions, but also of other people's. Put a basketball player in front of a screen showing a video of someone on a court, throwing a ball, and they're pretty good at telling you where that ball is going to end up.[22] Do the same thing with a journalist, say, with minimal motor experience of basketball, and they're not so good. 'If something is not in your motor vocabulary, you can still make predictions, but they are more purely visual, statistical predictions,' Keysers says. This explains why, even for a highly talented basketball player, the more they play, the better they get at predicting the outcomes of other players' physical movements – and the better at the game they get.

Now think back a moment to Yoko Ichino's proprioception work at Northern Ballet. In many of her classes, the dancers are allowed to have their eyes open, but the curtains over the mirrors are closed. Unable to see their own bodies, all they can see are *her* precise, expert movements. Focusing on those, and how their bodies are feeling, rather than looking at themselves in the mirror, puts them 'in her shoes', and no doubt helps them to get better, faster.

An important caveat to all this is that no one has yet conclusively identified single mirror neurons in the premotor cortex of a human. However, there is evidence from brain imaging studies, for example, in support of the idea that we do have a mirror motor system, which allows us to get an intuitive feel for what someone else's body is doing.[23]

If a mirror system does allow us to get 'inside' other people's physical actions, might something similar apply to bodily perceptions and emotions, and empathy?

Gazzola and Keysers have used brain imaging to explore the idea. It's hard to induce a positive emotion, such as happiness, in a noisy, claustrophobic MRI scanner. So they started with disgust, which is much easier to provoke. Through an anaesthesia mask, they pumped in unpleasant scents, such as putyric acid, which smells like rancid butter. ('That worked really quite well. Though we had to take one subject out because he started vomiting,' Keysers remembers.) They found that, in particular, lower parts of the insula (which we know is important for emotion) were activated strongly by this disgusting smell *and* also when a participant witnessed disgust in others. The level of activity wasn't equal – it was perhaps three or four times stronger for the actual experience than for the vicarious one – but the pattern of brain activation was the same.[24]

The insula, of course, receives all kinds of sensory signals, including smell, taste and interoceptive data. A dorsal (upper) region of the insula can send movement signals to your stomach. One of the purposes of this circuit is to trigger vomiting when you perceive the signs of poison or tainted food – if your mouthful tastes rotten, say, or somebody else near you is vomiting. This same dorsal region also receives sensory signals from your viscera, including your stomach and intestines. 'For us, this is a very interesting region, as it doesn't harbour abstract representations of emotion but the *visceral* representations,' Keysers notes.

With the mirror system for movements, it's thought that you map other people's actions and the associated sensations – like the touch of a peanut shell, or a tennis racquet, in the hand – onto the neural circuits that you would normally use to induce and sense such an action yourself. And with the emotion of disgust, it seems, you represent someone else's disgust using the brain circuitry that you use to represent your own.[25] It follows that if you've never felt disgust, you can't represent, and so empathise with, somebody else's. If, as Gazzola and Keysers suspect, this type of mirror process extends to other emotions, this could explain why people with alexithymia lack empathy.

This work is also helpful for understanding why some people with autism have difficulty with empathy. Problems with empathy have even been considered to be a key feature of autism. Geoff Bird is scathing about this. 'There has been this perception that people with autism don't have empathy. And that's *rubbish*. You can see this immediately as soon as you meet some autistic people.'

In fact, he got into studying interoception through work on autism. Early in the 2010s, he started to link certain symptoms of autism – such as a lack of attention, or even an aversion, to other people's eyes – to alexithymia.

Then he turned his attention to claims about empathy. Studies in the field had produced conflicting findings. Some had concluded that, no, people with autism can't feel empathy, while others reported the opposite. Self-reports from people with autism were also mixed: while some said they really didn't 'get' empathy, others said they definitely did and their empathic feelings could be so strong as to be overwhelming.

Bird, along with other colleagues, including Uta Frith, a psychologist and autism specialist, wondered if the presence or absence of alexithymia might in fact be key. While perhaps one in ten of the general population has alexithymia, studies had suggested that for people with autism, the figure is more like half.

The team recruited adults with autism to come into the lab, with a loved one – which tended to be their mother, Bird recalls. 'It sounds really horrible but we gave them electric shocks.' The idea was to map the parts of the brain that respond when a person is themselves hurt. Then, while they watched, their loved one was shocked, again while their brain activity was measured. About half of the autistic volunteers showed brain responses that were the same as those observed in a group of people known to have neither autism nor alexithymia. However, about half were clearly alexithymic. And once their alexithymia was taken into account, there was no difference in the brain indicators of empathy for pain between the people with and without autism. 'That was a big thing for us,' Bird says. 'A lot of the interpersonal deficits in autism can be explained by alexithymia.'

That study was published in 2010.[26] Since then, Bird and colleagues have amassed further evidence that for people with autism who do have problems with empathy, it's their alexithymia, caused by difficulties with inner-sensing, that's responsible.[27] But why *should* interoceptive difficulties be relatively common in autism? As we'll get to in the next chapter, all kinds of sensory over- or under-sensitivities (and often both) have been observed in autistic people.

In a separate study, Keysers and colleagues scanned the brains of a group of autistic adults while they were shown photographs depicting various emotions. 'What we saw was a *hyper*-response in empathy-related regions,' he reports. This work suggested that – in this particular group – their empathy response was not impaired but heightened. Work in the USA, meanwhile, has found a hyper-activation in the amygdalas of people with autism when they experience direct gaze.[28] They seem to interpret gaze as an intense threat.

For Keysers, the reason at least some people with autism struggle with social interaction and avoid faces could be, then, not because they're uninterested in faces or other people, but because when they look at someone else they get too much information – *too much* stimulation, which they seek to avoid.

It's possible, Bird agrees, that for some autistic people with alexithymia, their alexithymia results not from an inability to produce or register the bodily sensations of emotion, but because they have trained themselves to subdue these sensations, because they are distressing. If the bodily signals of emotion are overwhelming, it could be beneficial to strive to ignore them – to try to learn *not* to feel.

What if you grow up in an environment that is so terrible that you are often overwhelmingly distressed? Lieke Nentjes at the University of Amsterdam had studied incarcerated criminal psychopaths, and found that they do relatively poorly at inner-sensing tests. She's also observed that what seems to distinguish at least some violent criminal psychopaths from 'successful' psychopaths, who don't commit awful crimes, is a traumatic childhood. 'One thing that struck me when I talked to them was their upbringing, or rather the lack of it – emotional abuse, sexual abuse, neglect, a lot of physical abuse,' she says. 'People with psychopathy are good at inferring what other people are thinking. But I've heard them say that emotion is no use to them; all they felt during their upbringing was fear.' So learning not to tune in to normal inner sensory signals could be an adaptation to a horrific environment? 'It needs a lot more research, but that could be an explanation,' Nentjes says.

Stephen certainly experienced a troubled childhood. His mother, who he now believes had undiagnosed post-natal depression, subjected him to emotional abuse, putting him down, withholding praise. When he was six, she set the house on fire in an apparent attempt to kill herself and her children. For the rest of his childhood, he was in and out of care. Learning not to feel could have been a survival mechanism for him, too. Bird does add, however, that there are plenty of people with alexithymia who didn't have a horrific childhood, and who don't ever remember being overwhelmed by emotions – but rather, simply not ever really feeling them.

The link between poor inner-sensing and emotional difficulties has led some researchers to explore training this type of sensing as a potential treatment for these difficulties. But as good inner-sensing is linked to all kinds of benefits – including, as we know from Chapter 11, more accurate 'gut feelings' as well as greater emotional wellbeing and empathy – there could be broader benefits for us all . . .

Sarah Garfinkel and Lisa Quadt at the University of Sussex have gathered preliminary evidence that inner-sensing training can help people who have trouble working out what emotions they might be feeling, and so who often feel anxious as a result. They hope that this training method might alleviate anxiety in autistic people, by reducing confusion about what their body is telling them.[29]

In a pilot study, Quadt gave a group of healthy student volunteers a series of heart beat counting tests, like the one I tried, but she also gave feedback each time. 'So it might be, you said you counted thirty-three heart beats, but actually it was forty-four,' she explains. Then she administered the other commonly used cardiac interoception task, in which a series of beeps are played, and you have to say whether they are in synch with your heart beat, or not. Again, Quadt gave the students feedback after each individual trial, informing them whether they were right or wrong. Then, she sent them off to do some exercise – like star jumps, or walking up a steep hill near the lab – to get their heart pumping, which makes it easier to count heart beats. After that, they returned to run back through the tasks, again with feedback.

Each training session took about thirty minutes, and the volunteers did it twice a week for six weeks. But by just three weeks into the study, almost all were showing significant improvements in interoceptive accuracy, and also interoceptive awareness. The anxiety scores that Quadt was gathering at each session also showed a 10 to 12 per cent drop. This was a group of healthy student volunteers, without any diagnosed psychological disorders. The training made a difference to even these everyday levels of anxiety.

The team is now running a trial on 120 autistic adults, to see whether the training will work for them, too. But perhaps it could help anyone who feels they struggle to understand their own emotions. So is there an easy way for someone to try to train themselves?

'Ah,' Quadt says, smiling. 'Well, we are working on an app . . .'
In the meantime, this is what she recommends:

1. Sit somewhere quiet and set a timer (on your phone, watch or home digital assistant, perhaps) for one minute, but don't start it yet.
2. Close your eyes and try to sense your heart beat.
3. Now start the timer, and try to count your heart beats.
4. Do this again, but feel for your pulse this time, to take an accurate measure (this is the feedback that should help your interoceptive awareness to improve).
5. Repeat all the steps.

If you can't sense your heart beating, perhaps try some exercise first, because this makes it easier.

While problems with empathy are fundamental to alxeithymia rather than autism, alexithymia is still a lot more common in people with autism than people without it. Other disorders, including schizophrenia, also feature emotional 'flatness'. Problems with inner-sensing may help to explain these emotion-related symptoms, too.

In fact, a consistent finding, time and again, of inner-sensing difficulties, either in a person's accuracy or awareness of how accurate they are, or both, in a wide range of disorders – including schizophrenia, autism, eating disorders and depression, not to mention clinical psychopathy – makes Geoff Bird wonder if deficits in inner-sensing might represent the long-elusive explanation for a key mystery in psychology: why half of people who meet the diagnostic criteria for one psychopathology will also meet the criteria for a second, and half of the people who meet the criteria for two will also meet the criteria for three. There appears to be a general underlying risk factor for developing a psychopathology of some form, dubbed the P-factor – but no one's been able to nail its identity. 'There's a good chance it's interoception,' Bird thinks.

People with alexithymia not only have difficulties with emotion and empathy, they also don't sleep well, Bird and his team have found. He thinks this could be because they're not sensing the physical, muscular signals of fatigue. Chronic poor sleep, in turn, causes all kinds of emotional, cognitive and physical health problems.

So if poor inner-sensing interferes with your emotion processing and your relationships with other people, and your sleep – and so your ability to learn and to make sound decisions, and even, as the evidence suggests, your perception of having a single, continuous self[30] – it could plausibly be core to a variety of diagnoses. It's also been observed, Bird adds, that problems with inner-sensing can emerge during puberty, pregnancy and the menopause. These are all stages of life linked to the emergence of a variety of psychological problems and disorders, including anxiety disorders, depression and schizophrenia (which often develops in young adulthood).[31]

'This is really new stuff,' Bird says, with a mixture of caution and enthusiasm. 'But if we can start to say that yes, interoception has diffuse effects on all these things that are implicated in a number of different psychiatric conditions *and* we know that interoception has problems in puberty, pregnancy and the menopause, then we are starting to get a mechanistic idea about where psychiatric illnesses come from.'

It sounds exciting, I tell him. He nods. 'It *is* exciting.' He pauses, and shakes his head a little. 'It might be complete crap . . . But it might be really, really good.'

When we think of our emotions – joy, fear, anger, love, and so on – pain typically wouldn't feature in that list. But recall from Chapter 10 that pain has an important emotional component. When something happens to a loved one, we might feel that we're sharing in the pain that they're going through. Recent studies have shown that, to an extent, this is literally true.

Keysers and Gazzola, and, independently, the social neuroscience researcher Tania Singer, have all found evidence that when you feel someone else's pain, some of the same brain regions that are involved in your own pain perceptions are activated.

In a widely-cited study, published in 2004, Singer and her colleagues put members of sixteen romantic couples into an MRI scanner. When one person was given an electric shock, the team noted, as expected, activity in brain regions associated with both the discriminative and also the emotional aspects of pain. When a member of a couple was shown an image of their partner being shocked, while the discriminatory regions didn't respond, the emotional pain regions certainly did. The anterior insula, in particular, lit up.

Since then, there's been a lot more work confirming that this 'empathy for pain' network exists and that it doesn't distinguish between someone else's physical or psychological suffering.[32] 'The basic principle is the same,' Singer notes.

What this work shows is that to feel with another person, we need, at a neurological level, almost to blend ourselves with them – to dissolve, at least a little, the distinction that we maintain between 'self' and 'other'. Some people, like Stephen, don't blend very easily, if at all. But others fall further towards the other end of this spectrum.

In 2017, I was asked to write a news story about new research on something called the 'vicarious perception of pain', or sometimes 'mirror pain'. The paper described how about 27 per cent of people who watch, hear or are even told about someone being physically hurt feel either an acute flash of pain in the same body part, or more of a general, whole-body, nauseating type of pain.[33] Only *27 per cent*? Until I read that paper, I had no idea that not everybody did.

My own vicarious pain perceptions are very transient and they don't hurt, exactly. They're more like radiating electrical bursts that vanish after a moment or two. Sometimes I'll feel them in the same region as the injured person, but often, they happen in my legs.

Hugo Critchley and Jamie Ward, the synaesthesia expert from Chapter 1, were two of the co-authors of the paper. Because stimulation of one sense (vision or hearing) triggers perceptions in another (pain), vicarious pain is sometimes called a type of synaesthesia, but it's better viewed as a failure to distinguish properly between sensations that you yourself are experiencing, and ones that you're only simulating.

Why should we simulate pain to such an extent – in some cases right down to the spot where it's being inflicted? Perhaps, it's been suggested, because such localised representations help us learn how to react to a potentially damaging situation. When Alessio Avenanti at the University of Bologna and colleagues showed adults videos of people receiving an injection into the hand, the team noted an inhibition in the motor neurons that activate muscles in the viewer's own hands. Presumably, this was because we learn in childhood to keep an injured hand still, in a bid to prevent further damage. This type of finding suggests that we all have a complex mirror-like pain system, which represents more than just the emotional dimension of pain.[34]

However, people who actually feel mirror pain do show some brain differences to people who don't. They – we, I suppose – have more grey matter (relating to greater activity) in the insula and also the somatosensory cortex, and less in a region called the right temporoparietal junction (rTPJ), which is involved in representations of 'self' versus 'other'.

When the 2017 study team scanned the brains of the mirror pain group using fMRI, they found more communication between the anterior region of the insula and the rTPJ. Their conclusion? Such people 'systematically fail to attribute shared bodily representations to others'. In other words, we don't properly recognise our simulations as mimicry, and think that what's happening to someone else is happening to us.

Another member of this research team was Michael Banissy, a professor of psychology at Goldsmiths, University of London. Banissy is an international authority on a related phenomenon, often called 'mirror touch synaesthesia' (though again, I think the term 'synaesthesia' is misleading). Christian Keysers has found that when we see someone else being touched, we tend to activate similar parts of the brain as those that become active when we ourselves are touched. For people with mirror touch, this response is 'hyper-activated', Banissy explains, so that they actually perceive touch on their own skin. These people also show a greater tendency to imitate others, he adds, and are more vulnerable to the rubber hand illusion. Overall, 'they struggle with boosting the "self" and inhibiting the "other".'[35]

Bannissy's research suggests that about 1.6 per cent of the general population experience mirror touch.[36] That equates to more than one million people in the UK alone. Fiona Torrance, the multi-synaesthete whom we met in Chapter 1, also experiences both mirror touch and mirror pain. In fact, she reports not merely a blurring but an extraordinary dissolution of the boundary between her own sensations and other people's.

Now in her forties, Fiona remembers from a very young age feeling her conscious self sometimes slipping 'into' other people, or even animals. As a child growing up in South Africa, whenever she saw dragonflies, mosquitoes, butterflies and birds flying past, 'I would feel as if it was my body flying.' When she watched her parents hug, 'if I focused on my mother, I would feel that I *was* my mother being hugged. If I focused on my father, I'd feel that I *was* my father being hugged.' At one point, as we're talking, I inadvertently touch my chin with my pen and her fingers go to her own chin. She felt it, she explains, exactly as if my pen had touched her.

For Fiona, this feeling of entering another body has brought some advantages. As a child, she didn't have to practise to learn to play the guitar, the way most people have to. This wasn't because of her sound-to-colour synaesthesia, or at least, that wasn't the main reason. 'In fact, I was talking to my mother about this the other day,' she tells me. 'She said, "Your father never put your fingers on the strings. You could always play a tune by ear." But I remember, if I watched my father, I could feel that I *was* him, and I could see in my mind's eye where his – my – fingers were on the strings. That's how I learned to play.' This sounds like a version of one monkey learning how to deal with a peanut by watching and simulating the movements of another, but for a task that is vastly more complex.

Her hyper-simulations do not end there. While my own mirror pain is mild and fleeting, Fiona's couldn't be described that way . . . A friend once invited her to go to *The Girl with the Dragon Tattoo* at the cinema. She found the experience horrific: 'When the character Salander was tortured, I felt as if my body was being beaten.'

Emotional pain, too, is very hard for her to witness. If she sees someone in agony, she feels choked in her chest. This process of picking up on pain and other emotions happens to her automatically, all the time, even when just she's walking past people in the street.[37] 'I think that's what my mother found hard to cope with when I was a young child,' she tells me. 'That I'd have all these memories of the emotions of just passers-by.'

For Fiona, some of the clearest cues to how somebody else is feeling come from the colour of a 'cloud' that she sees around them, mostly around their hands and feet. For her, blue indicates pain. Orange, or sometimes a greyness, signifies sickness. A purply-black signifies anger – though if a person who's typically friendly becomes angry, their usual green will darken. (Her father's colour used to go from an olive green to a forest green when he was cross.)

Fiona says that she sees these colours around animals, too. She has a menagerie of pets and tells me she can read how they are from their colours. As a child, she lived on a farm, and she had a collie dog. 'One day, even though she had no signs of illness, I saw orange in her green, and had a feeling she was sick. A few days later, she became obviously sick from ingesting poison that had been sprayed on the farm. I rushed her to the vet. After a week in care, she survived.'

Coloured auras . . . An ability to use them as indicators of emotions and illness . . .These are remarkable claims. If you're dubious, I do understand. But while Jamie Ward isn't sure exactly what's underlying Fiona's 'emotion-to-colour' synaesthesia, he's convinced that there's something to it – she's not making it up, this is her experience – and there may well be something meaningful in it.

While it can be difficult for her to be around people in pain, Fiona reports that these experiences give her intense and immediate intuitions about other people's emotional and physical health. Modern research can go some way (if not all the way) towards explaining how this might happen. In a different time, and perhaps a different place, it's easy to imagine how someone like Fiona might have been revered as a traditional healer – or put on trial as a witch.

In fact, people without emotion-to-colour synaesthesia, or indeed any synaesthesia, but with sensory sensitivities can also report experiencing intense empathy, as well as feeling in rawer contact with the world. And to some researchers, this is intriguing. Could a distinct sensory processing profile help to explain the personalities of people we might otherwise just think of as 'sensitive'? And what about for other people? Could duller senses mean a 'thicker skin'?

In the next chapter, I look at the idea that distinct patterns of functioning of a range of our many senses are intimately linked with our personalities. Because it's become clear that to understand yourself, and your family and friends – to know *who* they are – you have to fully appreciate *how* they sense.

15

Feeling Sensitive

What being a 'sensitive' person really means

Think about a few of your friends – or your children, if you have them. Which adjectives would you use to describe them? 'Adventurous', perhaps, for one, or 'robust'? Maybe 'thin-skinned', for another; *'sensitive'*, even.

We're all familiar with the concept of a 'sensitive' person – the kind who feels jibes deeply, who readily cries at sad (or even happy) stories, and who is the stereotypical 'wallflower' at parties. But new research is revealing that these *personality* attributes are tightly linked to greater *sensory* sensitivity.

In fact, all kinds of studies are now linking how we sense with how we behave.

As we'll discover, for people at the sensory extremes, life can be very difficult indeed. But a better understanding of the role of sensory differences in explaining certain disorders – including autism and attention deficit hyperactivity disorder (ADHD), as well as Sensory Processing Disorder (a little-known disorder that, according to some estimates, affects one in twenty of us – so perhaps one child in every classroom) – holds the promise of revealing the role of our sensory settings in all our lives.

*

- Other people's moods affect me.
- I am easily overwhelmed by things such as bright lights, coarse fabrics or sirens close by.
- I seem to be aware of subtleties in my environment.
- I startle easily.

- Being very hungry creates a strong reaction in me, disrupting my concentration or mood.
- I make it a high priority to arrange my life to avoid upsetting or overwhelming situations.

These six items are taken from a twenty-seven-point 'Highly Sensitive Person' (HSP) checklist.[1] If you agree with fourteen or more, you may be an HSP. When I meet Elaine Aron, the American psychologist behind the concept and the checklist, and ask how many would apply to her, she smiles: 'I think I'd probably tick every one.'

Now in her seventies, Aron has spent decades developing the concept of high sensitivity as a personality trait, and one that's underpinned by general sensory processing sensitivity. Her pioneering work in this field was driven by her own experiences. As a child, she felt she was different to the majority. 'I think it was probably mostly that I did not seem to function well in groups of excited girls,' she tells me. 'I tried – but I just couldn't do it. In fact, I recently found a report card from kindergarten in which the teacher wrote, "Elaine is a very sensitive quiet child."' She gives a gentle laugh: 'So somebody else was seeing it too.'

At college, at the University of California, Berkeley, things didn't get much better. She'd sometimes find herself crying in the loos, struggling with the practical and social pressures of student life. It was only when a therapist suggested, in 1990, that she was 'highly sensitive' that Aron started to wonder about what that might mean scientifically. Her first step was to interview other people who reported feeling similar to her. (These participants were recruited through the University of California, Santa Cruz and a local arts organisation.) 'I was just curious,' Aron says. 'I thought if I interview people who think they're highly sensitive, I'll be able to find out what it is.'

From these interviews, Aron identified sixty factors – mostly sensory and emotional items – that seemed to relate to being highly sensitive. Her husband, Arthur Aron (one of the psychologists behind the scary suspension bridge/safe bridge study), took these items and streamlined them statistically, and, together, they created the first HSP scale. When they were done, they were amazed, she says, at the variety of items that were linked (so that if someone said yes to one, they were likely to say yes to the others). Someone who was very sensitive to pain was also likely to be highly conscientious, deeply moved by the arts and music *and* have a tendency to pick up on subtleties in the environment, for example. But then she realised that the connection between these apparently disparate items was a great depth of processing.

People like her, she figured, are more affected by what's happening around them. Because they are 'thin-skinned', everything from music playing in a café to the slings and arrows of outrageous fortune hits them deeper, triggering a more intense response. Their sensory sensitivity, which is set to 'high', means they can become over-whelmed by sounds or lights, or other people's emotions. At a certain level, such stimulation is tolerable, though still affecting. (I met Aron at a waterfront café near her home in Tiburon, Marin County and, given the gentle music playing inside and the noise of the occasional passer-by or car outside, it took some time to hit on the ideal place to sit . . .) But in a hectic sensory storm – such as a freshman dorm, a boisterous party, or even an open-plan office – without plenty of down-time, it can become very difficult indeed. 'The things people notice are the surface behaviour of highly sensitive people, such as being upset by too much noise or crying easily or not liking to be rushed on a decision,' Aron notes. 'But the most important aspect is that they process information more deeply and thoroughly.'

Her next research step after the interviews was to conduct a phone poll of 299 people living in Santa Cruz County. These randomly-selected residents were interviewed and given a short form of the HSP scale. Based on the results, Aron came up with an initial estimate that perhaps 20 per cent of people are highly sensitive – which of course means that 80 per cent are not.[2] Three hundred people is not a lot for a study like this. And these were residents of a relatively wealthy area, home to a major university; they weren't exactly representative of the US as a whole, never mind the rest of the planet. However, since that phone poll study, at least a few thousand people have been surveyed, Aron says, and a rough division of one-fifth highly sensitive and four-fifths not has been noted repeatedly.

Within the HSP population, there seems to be a spectrum of sensitivity, but everyone else falls outside it. Most of us, she argues, are simply not 'sensitive', and the difference between the two groups is 'as big as gender'. 'How it impacts people is *enormous,*' she says.

A clear division between a minority of sensitives and a majority that aren't might still sound like a pretty big call, based on limited research. But support for it has subsequently come from studies on everything from babies to a colourful little American sunfish.

In 1993, just a few years after Aron's phone poll, David Sloan Wilson, at Binghamton University, New York State, led one of the first systematic investigations into whether something equivalent to personality types might exist in animals. He and his team collected adult pumpkinseed sunfish from lakes located close to Cornell University, about an hour's drive away. When their offspring grew to be juveniles, the young fish were moved into an experimental pond at Cornell, for close observation.

The researchers quickly noticed clear and consistent differences between individual fish. After they put a novel object – a cylindrical minnow trap – into the pond, some of the fish lost no time in exploring it. These same fish were also less likely to take care to swim close to other sunfish, and more likely to approach a researcher who clambered into the water. When these individuals were then shifted to the lab, they got used to their tank pretty quickly. 'All of these differences indicated a degree of fearlessness,' the researchers noted.

In contrast, other fish were wary of the traps, and they showed another cluster of consistencies: in the pond, they swam close to other fish, they tended to avoid open water and they were more likely to flee a researcher who invaded their habitat. When moved to the lab, they took longer than the others to settle down.[3]

The team concluded that there are 'shy' and 'bold' pumpkinseed sunfish. And a fish's type had implications not only for themselves but also for other species in the ecosystem. When they were living in the pond, the bold fish gobbled up three times as many copepods (a type of small crustacean) as the shy fish. Going for copepods is riskier than going for weed-dwelling Daphnia, because copepods tend to be found in open water, where it's easier for a hungry bird or another, bigger fish to grab a sunfish. Though the bold and timid fish were of the same species, they had a notably different diet, as well as contrasting behaviour.

One of the reasons these results were important is, as the researchers write: 'Although everyone that works with animals knows that they have different personalities, the nature of individual differences in a population has seldom been the focus of study.' Also, this work revealed a clear shy–bold division. But why *should* there be this personality split among sunfish? Why, for that matter, should the same dichotomy go on to be noted in subsequent studies on everything from goats to great tits to pigs?[4]

In 2019, I caught up with David Sloan Wilson at a conference. A distinguished evolutionary biologist, he was at the conference to talk about fascinating new work on human groups. But I really wanted to ask him what he made of Elaine Aron's ideas, and her referencing of apparently related animal research, including his on the sunfish.

He was immediately enthusiastic. 'I love that work! It's a great idea and it does appear to exist in other species. There's a video of it in pigs. Have you seen it?' I shook my head. 'So, you have pigs running along a route and turning left at the end, to find food. They all learn to do this. They go up, turn left, and get the food. Then, you put a novel obstacle – a pail – at the top, right before they turn. Some will just ignore it, barrel on and turn left. Others will stop in their tracks, spend a minute looking at it, all very timidly. These are the highly sensitive ones – think of Wilbur the Pig in *Charlotte's Web*. Then, you move the food to the right. The ones that stopped at the pail quickly learn to go right to get food. The others – the inattentive pigs – keep going the other way. They take a lot longer to learn to go right.'

Over and again, some animals within a group have been rated as brasher – and as risk-takers – while others are more responsive, reactive, flexible and sensitive to their environment. Jaap Koolhaas at the University of Groningen led a review of work into differences in aggressiveness in several species of bird and mammal, and suggested that the fundamental difference between 'aggressive' and 'non-aggressive' individuals was really one of relative insensitivity versus sensitivity to the environment. In some bird species, 'aggressive' males were observed to develop routines quickly and to be more rigid in their behaviour, while members of the 'non-aggressive' group were more flexible and responsive to sensory information about what was happening around them.[5]

Aron thinks that these two ways of being reflect distinct survival strategies.[6] If times are tough, a brasher bird might fight over food. A more cautious, observant bird might instead recall where it once glimpsed a distant tree with a decent crop of berries, and fly off there instead.

'Sensitive' animals and people, who are more reactive to all kinds of sensory signals, will take longer to observe and respond less rapidly. This can make them appear more timid, less impulsive and also more risk-averse (though, Aron stresses, when they encounter a familiar situation that offers an opportunity others have missed, they can move very fast indeed). In contrast, animals and people that are *less* responsive to these sensory signals function more on automatic pilot, which can make them quicker to rush in.

The sunfish work was itself inspired by research on children. In the 1950s, two psychologists, Stella Chess and Alexander Thomas, pioneered the study of temperament in infants – how, from their very earliest days, children seem to show distinctive patterns of behaviour, with some being easy-going, while others are often crotchety or fussy.[7] At the time, most psychologists thought that personality was determined by experience; and after the Second World War, there was a fear of any discussion of genetic differences. But, as Aron puts it, 'Chess and Thomas saw past all of this, at what every parent or teacher knows – children differ.'

In the 1980s and 1990s, Harvard psychologists Jerome Kagan and Nancy Snidman further developed this research and characterised a temperament trait that looks very much like high sensitivity. Kagan and Snidman first established not only that there are clear differences in the temperaments of babies, but that these differences tend to persist. This work began in 1986, with 500 four-month-olds. When these babies were presented with a series of colourful new toys, about 20 per cent consistently thrashed their arms and legs and cried, 40 per cent were unfazed, and the rest were somewhere in between.

As these children grew up, they came back again and again for further tests and interviews.[8] At the age of eleven, 20 per cent of those initial 'reactive' screamers were classed as shy when interviewed by the researchers, and they also had a stronger physiological response to stress. Of the chilled-out babies, about a third had grown to be confident and sociable. Most of the kids had drifted towards a kind of middle ground of reactivity. Only 5 per cent of the chilled 'un-reactive' babies had switched to become 'reactive' pre-teens, and the same figure held for reactive-to-chilled. Environment – in terms of home and school life – clearly had an impact. But this work has been taken as showing that there's a genetic contribution to temperament in babies, and to personality in early adolescence.[9]

Thomas Boyce, an emeritus professor of paediatrics and psychiatry at the University of California, San Francisco, has spent over forty years working with children. His research has led him to classify the majority of kids as 'dandelions'. Like this plant, they're resilient, and as long as their environment isn't terrible, they'll do pretty well anywhere. ('Put them in a tough or a good environment and they'll basically flatline – they'll do the same,' says David Sloan Wilson, who is a fan of this work, too.)

But about 20 per cent – the same proportion that Kagan and Snidman identified as 'reactive' as babies and that Elaine Aron classes as highly sensitive adults – are what he terms 'orchids'. Orchids are more sensitive to their surroundings. In a neglectful, abusive environment, they'll suffer badly. But in a warm, nurturing environment, they'll flourish.

Boyce has explored this in the lab. He's looked at how children respond in all kinds of situations, from talking with a stranger to having a drop of lemon juice put on their tongues. He's found that while some kids – the ones he dubs orchids – have a strong fight or flight response and experience a notable increase in levels of the stress hormone cortisol, for example, the others have a much more muted physiological reaction.[10]

High reactivity – or being an orchid, in Boyce's terminology – has been linked in various studies to a particular version of a gene involved in regulating levels of the neurotransmitter serotonin.[11] A study of children who were reared in Romanian orphanages during the horrifically negligent period before the fall of Ceauşescu, in 1989, and who were adopted by the age of eleven, found that, on adoption, those with this short version of the gene had the highest level of emotional problems. Among this group, those who then suffered a relatively high number of stressful events had the highest overall emotional problem scores at fifteen. However, those who experienced very few stressful events in this period went on to show the biggest *drop* in emotional problem scores. The more vulnerable kids who were moved to a supportive, relatively stress-free home flourished, as Boyce would have predicted.[12]

Elaine Aron thinks that orchids probably grow up to be HSPs, and Boyce is certainly receptive to the idea. 'What she sees in her primarily adult patients is very close to what we are seeing in orchid children,' he says.

Michael Pleuss, a psychologist at Queen Mary, University of London, does research in the same area. He prefers the term 'environmental sensitivity'. For him, environmentally sensitive people are better able to register and process signals from their environment – whether those signals are good (loving parents or exposure to music and visual art, perhaps) or bad (such as neglect or a natural disaster).

In 2018, a team led by Pleuss, which included both the Arons, published findings on British children using a new Highly Sensitive Child scale. This twelve-item questionnaire, which drew on the Arons' HSP scale for adults, was designed for children aged eight to nineteen. It taps into purely sensory processing experiences (for example, 'I don't like loud noises', 'I love nice tastes'); observation of subtleties ('I notice it when small things have changed in my environment'); and also feeling overwhelmed (for example, 'I find it unpleasant to have a lot going on at once').

Tests on separate groups of kids consistently identified three groups: about 20 to 35 per cent were highly sensitive, 41 to 47 per cent were of medium sensitivity and between 25 and 35 per cent showed low sensitivity.[13] This does seem to map on to Kagan's three-group findings on children.

What might it mean for Boyce's model? Pleuss argues that perhaps a third, moderately-sensitive 'tulip' group has been hiding in the personality bouquet.[14]

Whether you call kids who fall towards one extreme end of the spectrum orchids or highly sensitive or environmentally sensitive or reactive, it seems clear that the way other people handle them can make a huge difference to their psychological wellbeing not just in their teenage years but into adulthood.

Aron has published work finding that adults who scored highly on the HSP scale and who'd had a troubled childhood also had especially high levels of emotional distress.

Remember, Elaine Aron says, that while she would tick yes to all twenty-seven items on her HSP checklist, there are people who would tick no to every single one: Fireworks? *Bring them on*. Noisy parties? *Love them*. People at these extremes live in essentially two different worlds, she argues. There's no point insisting to a highly sensitive child that being close to a fireworks display is fun, just as there's no point telling a friend who sees #thedress as blue and black that it's white and gold. What you perceive as being the case simply *is* the case for you.

If a highly sensitive child has a parent or a teacher who is markedly not sensitive, this can be tough – and for the parent or teacher, too, Aron says. 'The thing about raising a sensitive child is you want to balance getting them to try new things that they are hesitant about and yet not pushing them to the point where the situation's traumatic or where they feel bad about themselves. You really have to listen to the child.'[15]

Carie Little Hersh, an associate professor of anthropology at Northeastern University, Boston, US, writes a blog about the relevance of anthropology in everyday life. In one post, she writes about reading one of Elaine Aron's books, *The Highly Sensitive Person*. 'It was like my own personal *Da Vinci Code* – riveting, compelling and totally solved a mystery about myself I didn't know existed. My whole life I felt more worn out than others, more overwhelmed and more over-stimulated. But being raised in a Catholic family with a Protestant work ethic and an American intolerance for anything perceived as weak, I saw my sensitivity and heightened perceptivity as personal failings to overcome.'[16]

Hersh poses some of her own self-referential questions – and the types of responses she'd become used to:

- Why do sore muscles, or tight shoes, or pokey car headrests make me so irritable after ten minutes that I'd do anything to end the sensation? *Take some ibuprofen and toughen up.*
- Why, despite my curiosity and interest, can I not last more than twenty minutes in a loud bar or street festival before I'm sprinting home? *Stop being so whiny and just relax.*
- Why am I completely tapped out and overwhelmed by 5pm? *You're just being lazy.*

Though in some ways, her kids have different personalities, they are also sensitive, she writes. They hate anything sticky on their hands, they dislike getting wet, they refuse to wear any non-fleece material, they can quickly sense other kids' emotional states. Neither she nor they like change.

In Aron's long experience, the highly sensitive people who flourish are indeed those who grew up in a supportive environment – the orchids who got the person equivalent of a medium-grain bark mix, warmth and plenty of light. Depending on where you live in the world, it can be harder, though, for one group of HSPs in particular to get those ideal growing conditions in childhood and on into adulthood.

'Sensitive men are very different to other men,' Aron observes. That can be a problem if you grow up in a culture that doesn't view sensory sensitivities and strong emotional responses as masculine. 'It's a big thing for them, because people associate sensitive behaviour with femininity – so it can be easy to think they are more feminine. But they're not.'

Ted Zeff, a psychologist who was based in the San Francisco Bay area, specialised in therapy for highly sensitive boys and men. A strong supporter of Elaine Aron's work, as she is of his, Zeff surveyed men in various countries to explore how the trait of high sensitivity is regarded outside North America. Zeff died in 2019, but I managed to talk to him about his work before he passed. He stressed that the 'skewed view' that there's something wrong with highly sensitive men is culturally based. 'One subject in my research, who was from

Thailand and who is a highly sensitive man, told me that since kindness and sensitivity are highly regarded in Thailand, he was elected president of his class in school,' he said. 'The men I interviewed from India, Thailand and Denmark stated that they were rarely discriminated against for being sensitive – while the men from North America were frequently discriminated against.' Other studies back this up. A survey of attitudes in China and Canada found that while in China, 'shy' and 'sensitive' children were generally seen as desirable playmates, in Canada, they were not.[17]

Historically, the HSPs in a society were the priestly advisors and shamans, Zeff believed. In the modern world, Aron has found that they are over-represented in some careers – particularly in self-employed, creative professions. (The singer-songwriter Alanis Morrisette is just one self-acknowledged HSP who has become a friend of Aron's, via the research.) If sensory signals have a bigger impact on you, they will grab and occupy your attention. It makes sense, then, that someone who is more responsive to musical notes could become more engaged by them, or that someone captivated by the colours of a garden might want to try to paint it.

Sensory sensitivities are, as we know, common among synaesthetes, who are also over-represented in creative fields – with one notable exception: high-end cooking. Idiosyncratic associations between music and colour may be pleasing to non-synaesthetes too – the paintings of Kandinsky, a famous synaesthete, are a case in point – but, observe Charles Spence, chef Jozef Youssef and University College London senses researcher Ophelia Deroy in a recent paper: 'synaesthetic creativity, when expressed in edible form, may simply not be very tasty.'[18]

It's also easy to understand how someone who, because of their sensory sensitivities, perceives things that other people don't might struggle to explain these perceptions. A couple of summers ago, I got an email from a former public affairs officer based in Washington, DC, called Mike Jawer. I'd recently written a feature about sensory sensitivity. He'd read it, and he wanted to ask whether any of the people I'd interviewed had mentioned paranormal-type experiences. He was asking, he explained, because he himself had interviewed a bunch of HSPs, and many had shared extraordinary stories with him.

Jawer's interest in this started while he was working on a US Environmental Protection Agency investigation into which features of certain buildings or office environments might cause sick building syndrome, in which some people, but not others, report various health complaints from being in a certain building, often their workplace. Jawer talked to all kinds of people, and soon realised that there was a common thread linking workers who reported headaches or respiratory problems in particular buildings compared with those who didn't: they also tended to report various sensory sensitivities. Some among this group then went on to tell him that they had other kinds of unusual experiences . . . they saw ghosts, perceived auras or felt presences. This led Jawer to survey a broader group of HSPs. Again, he noted a preponderance of anomalous perceptions among them.

Jawer and I don't interpret his findings in the same way. He has a strongly spiritual take on it all (feeling that more sensitive people can detect genuinely supernatural phenomena), and I don't. But his open-minded curiosity led him to some intriguing observations, and if viewed through a sensory sensitivity lens, they make sense, don't they?

People who are highly sensitive across various types of sensing will hear, smell, feel things that others do not. If other people in precisely the same environment can't detect them, and therefore deny that they are real, this could lead the sensitive person towards notions of the supernatural.

Research that potentially supports this idea came from an unusual study of concert-goers, in 2003. For this work, the British composer Sarah Angliss helped to devise psychological experiments that incorporated 'Soundless Music' – infrasound far below the frequencies that most of us can sense with our ears, and so can't hear, but which can stimulate mechanoreceptors elsewhere in the body.[19] As Angliss puts it, infrasound 'laces sacred organ music in cathedrals around the UK'. It's been suggested that it contributes to feelings of awe, especially in the context of a religious service.

Not knowing what they were getting themselves into, 750 oblivious concert-goers turned up at the Purcell Room in central London to listen to pieces by Philip Glass, Debussy, Angliss, and others.

Though they weren't aware of it, some of the pieces, but not all, were dosed with infrasound, which was created and carefully controlled using a generator run by acoustics specialists from the UK's National Physical Laboratory. At the end of the concert, the audience members completed a questionnaire that asked whether they'd had 'unusual experiences' during any of the individual pieces, and if so, to describe them.

Overall, during pieces that included infrasound, reports of such experiences were 22 per cent higher. Respondents wrote about feeling a 'shivering on the wrist', an 'odd feeling' in the stomach, a rise in heart rate, feeling anxious, and even sudden memories of emotional loss.[20] Not all the concert-goers had these experiences, implying that some were more responsive to the infrasound than others. Infrasound is caused naturally by heavy traffic, engines, even fans in AC units. In other contexts, it has been linked to reports of ghost hauntings, and especially a 'feeling of presence' – at least for some people.

As Richard Wiseman at the University of Hertfordshire, one of two psychologists on the Soundless Music project (the other was the ghost researcher Ciarán O'Keeffe, mentioned in Chapter 10), told the *Guardian* newspaper at the time, expectation, based on where you are, could easily influence how infrasound perceptions are attributed: 'If you walked into a modern building and suddenly felt sort of ill but didn't know why, it might be sick building syndrome. If you walk into an old Scottish castle with a reputation, that's a ghost.'

If you have sensory sensitivities and you score highly on the personality trait of 'openness', the odds that you'll see a ghost may be even higher. 'Tolerance of ambiguity, emotional ambivalence and perceptual synaesthesia' are all characteristics of the open person. And more open people are also more likely to believe in paranormal phenomena.[21]

The trait of 'high sensitivity' is not included in any of the commonly-used models of personality, including the wildly popular five-factor model. As we know, this model defines your personality according to your scores for extraversion, neuroticism, openness to experience, agreeableness and conscientiousness.

There are some problems with this model, however. Not least that psychologists disagree about even what 'extraversion' means, exactly. Also, what is meant by 'extraverted' and 'introverted' has changed with time. When the Swiss psychiatrist Carl Jung defined these terms in 1921, he described people who were introverted as directing their 'psychic energy' inwards, towards their own thoughts and feelings, whereas extraverts directed their attention outwards, to the world around them. An introvert carefully weighs up options before reaching a decision, he explained, and seeks to avoid stress.[22] This sounds quite a lot like a highly sensitive person.

However, that's not what a low extraversion score means today. Introversion is now understood more in terms of being less responsive to social reward – to getting *less* out of being with other people. Naturally, this would influence how someone behaves. It's argued that because extraverts derive more pleasure from interacting with others, they are driven to be gregarious and also to engage in riskier behaviour, in a bid to impress other people. In contrast, introverts just don't feel the same need to engage.

It's certainly possible to be enthusiastic, ebullient and assertive – to be an extravert, according to a common definition – and to be highly sensitive. In fact, though Aron herself is both highly sensitive and introverted, she has found that about 30 per cent of HSPs are extraverts.[23]

Jacquelyn Strickland, a highly sensitive extravert and counsellor, has interviewed many HSPs over the years, and is keen to stress the difference between high sensitivity and introversion. The highly sensitive extravert needs to gain energy from the external world, she writes. But they also need time alone to rest and recover. 'After our physical and mental energies are recharged by being "in", we go "out" to manifest our visions, our passions, or our work in the world.'[24]

Whether you're an extravert or an introvert, if you are a highly sensitive person, just knowing that is important, Aron thinks – because then, as she puts it, you won't feel that you have to be like the other 80 per cent. Though she's done a great deal to spread the word about her research, through her writing, retreats and a film, *Sensitive and in Love*, she wants do more. Thinking back to her own experiences at university, and how so many of her freshman colleagues were clearly so different to her, she says, 'I would love to be able to get in front of college counsellors because I feel sure some HSPs drop out because they're overwhelmed . . . I'm sure some commit suicide.'

Depending on their circumstances, for people who score highly on the HSP scale, everyday life can certainly be a challenge. For some people, however, their sensory variations – either sharpened hyper-sensitivities or dulled hypo-sensitivities – are so extreme that they cross the line into representing a disorder. A highly sensitive person might need plenty of time at home to recover from a party. Someone with extreme sensory sensitivities can struggle even to leave the house.

Sensory over- and under-sensitivities are often present in people diagnosed with autism. They're also fundamental to Sensory Processing Disorder, which some researchers think represents the extreme end of the highly sensitive spectrum. While research over the past decade has transformed how this disorder is viewed, there is also now a shift in thinking about the role of atypical sensing in autism. For some leading autism researchers, these sensory difficulties are now moving from the periphery to centre stage.

As a baby, Jack Craven would not sleep unless he was being held by an adult who was sitting bolt upright, one hand pressed firmly on the top of his head. 'If we relaxed or moved our hand or nodded off, he would start screaming again,' his mother, Lori, remembers. 'We would take four-hour shifts. It was brutal, it really was.'

Later, when Jack started to walk, Lori remembers putting him on a shiny marble floor in Bloomingdale's, the famous New York department store. 'There were a lot of reflections, and his whole body started shaking. It looked like he was having a seizure.' From a young age, Jack also found it unbearable to be anywhere loud: 'There was a lot of screaming if it was noisy,' Lori recalls. 'Actually, there was just a lot of screaming from him . . .'

At the age of six, Jack started telling his parents that he wanted to die and that 'God made a mistake when he made me.' Lori says she didn't even know that kids his age could think such things, and adds, 'Can you imagine your child saying that?'

Now fifteen, Jack still has profound sensory sensitivities. For Lori, who until recently home-schooled him in Roswell, north of Atlanta, Georgia, part of the process of helping Jack has been to encourage him to appreciate and control what she tells him are his 'super-powers'. And, in fact, although Jack's sensitivities do make life difficult for him, as well as his parents and sister, he does have some exceptional abilities.

'There's a test you can do: you look at a picture then you look away and you remember all the details you can. Jack will remember *everything*,' Lori says. Not only is Jack's ability to absorb visual detail superb, but he has perfect pitch: 'Oh my God, Jack can sing! He also changes his voice – like, he can do both Lennon and McCartney in Beatles songs. And he can imitate accents very easily as well.'

When Jack develops a particular fascination, his total absorption can drive him to do things most kids the same age would not contemplate. The first time I talked to Lori, by Skype, when Jack was twelve, our conversation was interrupted at one point by a sudden banging at her end. She glanced up at her ceiling. 'You hear that? He buys Nerf guns at a thrift store and takes them apart and paints them and changes the spring, and makes them better and faster . . . I'm pretty sure he has about sixty Nerf guns up there.'

As many as 90 per cent of autistic people have sensory problems – often over-sensitivities – and Jack's attention to detail, whether in a visual scene or a soundscape, is characteristic of autism, too.[25] Periods of intense absorption, meanwhile, are a recognised if sometimes under-appreciated symptom of Attention Deficit Hyperactivity Disorder.

For a while, Lori and her husband were worried about both. But autism did not seem to fit. 'When he was there with you, he was *with* you – he was locked on,' Lori says. And though doctors did wonder about ADHD, medication that can be effective for kids with ADHD made no difference to Jack. In any case, Lori was cautious: 'It's a common thing when you have a boy for doctors to be dismissive and to say ah, it's ADHD, take this medication. But I didn't want to cover symptoms. I wanted to deal with whatever it was.'

Lori's search for someone who could help to get to the root of

Jack's problems led her to Elysa Marco, at that time a paediatric neurologist at the University of California, San Francisco. When Jack was eleven, Lori took him to meet her. Marco agreed that Jack does not have the social difficulties that are fundamental to autism. Rather, she established, he has Sensory Processing Disorder. Though Jack's ears and eyes and other sensory organs are completely typical, the way his brain handles incoming sensory information is not.

Elysa Marco is now regarded as a leading world expert on SPD. Go back twelve years, though, and she hadn't even heard of it. As an autism specialist, she was, however, starting to think more about the role of sensory problems in the symptoms of many of her young patients. In her consulting rooms, she was seeing kids with a range of brain-related difficulties. 'And what I realised was that the families were coming in, and I would want to talk about the kids' seizures or their headaches or their language problems in the case of, say, kids with autism,' she says. 'The parents wanted to talk about that also. But what they really wanted to talk about was the minute by minute, the day by day, which was so hard because they couldn't get their kids into the shower to wash their hair because the kids wouldn't let them touch their heads, or they couldn't get a shirt on them because they would scream bloody murder, or they couldn't make soup in the kitchen with the blender because the kid would cover their ears and run out the door.'

In the 1960s, Jean Ayres, an occupational therapist and educational psychologist working in California, first identified SPD (or sensory integration disorder, as she called it then) as something distinct. Some people with SPD are under-responsive, which leads them to crave the stimulation of that sense or senses, while many are over-responsive in one or two senses, or more. Some are over-responsive in some senses and under-responsive in others.

Lucy Jane Miller was one of Ayres's students, and has researched SPD for more than thirty years. Now a professor of paediatrics at Rocky Mountain University of Health Professions in Colorado, and the founder of the Sensory Processing Disorder Foundation, Miller has developed assessment scales for diagnosis, coordinated research into therapies and done all she can to spread the word that SPD exists. According to some studies, a form of it affects between 5 and 16 per cent of us.[26]

In the summer of 2008, Miller gave a presentation at the University of California, Davis Medical Investigation of Neurodevelopmental Disorders (MIND) Institute about problems with sensory processing. Elysa Marco was in the audience. 'It was as though a big bright light turned on for me,' Marco recalls. 'I was very excited. I thought: okay, this is the way I need to think about and study my kids.'

Marco stresses that sensory sensitivities are not uncommon in childhood – and they certainly do not necessarily mean that a child has a disorder. 'If you can take your child to the fireworks and they cover their ears and they make it through, then they go home and they are back to normal and everything is fine . . . then bring earplugs. But if you can't take them anywhere there might be a popping noise, or every time you vacuum the house they are screaming for hours on end, or you put diapers on them and they scream and claw at their skin, then you cross the line.'

In her consulting rooms, Marco has observed that many of the parents of the kids she sees with SPD report similar but weaker, less debilitating symptoms. Lori Craven is one.[27] 'I remember getting to the end of an assignment in fourth grade and recalling nothing, because there was a child next to me and he was tapping his pencil on his desk,' Lori tells me. 'And I remember trying to take a test and I could hear the scratching of the pencils and I could hear the ticking of the clock and I could hear the buzzing of the fluorescent lights and I just remember tears falling on the page . . . I was thinking, what is happening, how come everybody else can take this test and I just can't function? Also, I'd noticed at a pretty young age that when anything was particularly loud and startling everyone else got on, but it felt like a shock in my elbows of all things. I could feel something physical happening to me.'

Even now, Lori always carries earplugs with her. She'll use them when neighbours are mowing the lawn or when a baby is crying somewhere near her. 'No one likes the sound of a screaming child but I can't calm myself down. With earplugs, I can still hear everything but it just kind of takes the edge off it.' With proprioception, too, Lori recognises problems, but in this case, it's an under-responsivity.

In gym classes that required any coordination she was, she says, 'a total clown . . . I do yoga almost every day now and I've realised my proprioceptive processing is unbelievably awful.'

For Lori, these problems are 'issues', which she feels able to manage herself. She's never had a sensory-related diagnosis, but she knows first-hand how deeply distracting sensory over-stimulation can be. It's difficult to concentrate if you're acutely aware of the ticking of the clock or the scratching of pencils. Problems with attention are of course a symptom of ADHD, but *under*-responsivities may drive behaviour associated with both SPD and ADHD.

At a meeting on sensory processing problems in Chicago, I met Rachel Schneider, a friendly, vivacious young woman with SPD. We sat down to talk in a bland ante-room to the hall in which she would shortly be speaking. To write that it was hard for her to settle would be putting it extremely mildly. 'How am I feeling in this room right now? *Horrible*. This is a terrible room. I'm trying not to focus on the lights and let them bother me. I'm trying not to listen to that echo – because I'm hearing my voice in my throat, and in the air *and* bouncing off the walls. And we are sitting with this gap behind us, so I'm floating in the middle of the room and a piece of me is going, "I hope this doesn't screw me up for when I have to be on stage" . . . I guess the door is *locked*? No one's coming in, right?'

Clearly, Rachel has hyper-sensitivities to lights and to sounds. She's also an emotional 'sponge', she says. When someone walks into the room, she can immediately sense how they're feeling – and if they're feeling bad, this can leave her reeling.[28]

Jack Craven's mother, Lori, says the same is true for her son. When Lori booked a place to stay in San Francisco for Jack's visit to Elysa Marco, she hadn't realised it was in a rough part of town. 'As soon as we walked out of the hotel, Jack would grab our hands really tight and he was shaking, he was terrified,' Lori recalls. 'He said, "I don't like San Francisco! There are too many sad people!"' (At home, Lori, her husband and their daughter can never react with anything other than warmth to whatever Jack does or says, no matter how hurtful it is. If they react with disapproval, 'we've got an explosion on our hands'.)

However, Schneider goes on, she's under-responsive to touch and, like Lori Craven, to proprioceptive signals. She has a *need* for physical contact. 'I'm a hugger!' And she feels a compulsion to be fiddling with something – a necklace, perhaps. The best way to get proprioceptive information, she's found, is for her to apply pressure to her limbs, or to move. Sometimes, Rachel says, she feels she just has to get up and jump up and down.

Why should an under-responsivity to touch and to proprioceptive signals lead her to crave them? The predictive processing theory of perception can explain it. As we know, perception isn't just about passively interpreting sensory data. But when the data are unclear, we're motivated to gather more. As Anil Seth explains it: 'I can minimise sensory prediction error by either changing my predictions or changing the data – by moving, for example.'

Moving closer to a dimly-lit vase, say, would enhance the data available, allowing you to gather more detailed visual signals and so generate a more accurate perception – seeing that it features a tree pattern, perhaps, whereas before you'd seen human figures. For someone whose brain isn't getting enough in the way of proprioceptive signals, stimulating proprioceptors by pressing on muscles or moving limbs is the equivalent of getting closer to that vase. It helps to improve the accuracy of perceptions of the position of the various body parts in space.

Moving constantly in order to gain more data on the position of your body could be interpreted as hyperactivity, or at least trouble with keeping still and focusing on a task in hand. There can be overlap, then, in how Sensory Processing Disorder and Attention Deficit Hyperactivity present.[29] In fact, a national US study led by Lucy Jane Miller in which parents were asked about ADHD and SPD symptoms found that about 40 per cent of kids with symptoms of one had symptoms of both. (About one in six kids with ADHD is estimated to have sensory impairments that interfere with their everyday life.) And there's overlap with autism, too.

Leo Kanner, a Johns Hopkins psychiatrist, published the first paper on autism, in 1943.[30] In these early case studies, he noted sensory peculiarities. A few of the children whom he described really didn't like the vacuum cleaner, for example. It's long been known that sensory problems are very common in autism. But the core characteristic symptoms were (and still are) deficits in communicating and interacting with other people, and also restricted, repetitive patterns of behaviour, interests or activities – physical rocking, or an obsession with baseball, perhaps.

For a long time, research into autism focused on the social problems. In 2013, however, the 'psychiatrist's Bible', the *Diagnostic and Statistical Manual of Mental Disorders* (*DSM-V*), included sensory problems among the core symptoms:

> Hyper or hypo-reactivity to sensory input or unusual interests in sensory aspects of the environment (e.g. apparent indifference to pain/temperature, adverse response to specific sounds or textures, excessive smelling or touching of objects, visual fascination with lights or movement).

This did help to change things, says Marco: 'More of the clinical world has got on board with recognising that there are these sensory based behaviours, and that you need to think about how a child is taking in sensory information, and how that information feels and is affecting their behaviour.'

Caroline Robertson, head of the Dartmouth Autism Research Initiative at Dartmouth College, USA, and Simon Baron-Cohen, head of the Autism Research Centre at Cambridge University in the UK, co-authored a review of evidence in 2017, which led the pair to this closing position: 'sensory symptoms are core, primary characteristics of the neurobiology of autism.' This marks, they wrote, 'a revolutionary shift' in the conception of autism, compared with the early days of research.[31]

Neurobiological research led by Elysa Marco has also shown that autism is distinct from Sensory Processing Disorder. When Rachel Schneider talks to me about the publication of this work, her excitement level soars. She practically bangs her fist on the table. 'It was pivotal! PIVOTAL! When I first heard about it, I was so excited, I wanted to throw a parade!' She pauses for a moment. 'And I *don't* like parades.'

For Rachel, these papers were important mostly because they demonstrated a neurobiological basis for her everyday difficulties

– they were proof to others that these problems were real. Marco and her team had reported finding poorer 'wiring' between brain areas that handle basic sensory information in a group of children with SPD and also in a group diagnosed with autism. But there were some differences between the two groups, too. Only the kids with autism had weaknesses in connections important for social-emotional processing – in tracts linking the area that handles visual face processing and the amygdala (which spots threats), for example.[32]

Difficulties with eye contact and interpreting facial expressions are common features of autism. And Marco and her team found that the weaker the connections between the face/emotion processing regions, the greater the child's social difficulties.

Developing ways to characterise sensory differences in people with autism properly is ongoing. And part of the reason that sensory difficulties haven't received the attention they deserve as potential *causes* of social problems in people with autism is probably because, Caroline Robertson says, 'the evidence hasn't been strong enough . . . But I think it's there now.'[33]

Christian Keysers, the neuroscientist and empathy specialist we encountered in Chapter 11, has an interest in this. One difficulty with comparing the results of brain imaging studies on autistic people is that there's so much variation in the extent of the core problems. Some autistic people can't talk. Others are highly verbal. Kanner himself noted that while some autistic children forged some degree of social relationships, others never did.

So, Keysers and colleagues decided to gather data on the biggest group of autistic people that they could muster: 166 males aged between seven and fifty, and also 193 males without autism, for comparison. Rather than starting out with any assumptions about what differences they might find between the two groups, the team measured activity across the brain for fifteen minutes and let the brains 'speak'. Yes, potentially there might be a lot of variation between the autistic people in the study – but, Keysers and his colleagues wondered, what, if anything, might *unite* their brains?

One thing stood out. What was most unusual about the brains of people with autism was strongly increased connectivity between

the thalamus, which receives and relays incoming sensory information, and the primary sensory cortices; in this study, mostly the
primary somatosensory, auditory and visual cortices. What's more,
the greater this 'over-connectivity' in any one individual, the higher
the level of their autistic traits. Strong connectivity between the
thalamus and these sensory cortices suggests more in the way of
visual, sound and touch input; these incoming signals are having a
bigger impact.

Among the group without autism, though, there was an interesting
pattern: connectivity between these regions was strongest in the
children, and it grew weaker with age.[34]

To imagine or think about something, you need to be able to
uncouple your internal experience from outside stimuli. 'If you were
to close your eyes and think about the last time you played tennis,
if I was scanning your brain, I'd see that your sensory cortices were
now no longer under the control of your thalamus but your higher-
level cortices,' Keysers says. Normally, with age, our ability to do
this improves. We get better at holding at bay streams of signals from
our sensory receptors – at 'stepping outside' the present, sensory
moment. 'What we saw is that in the autistic brain, this is not the
case.' The team didn't set out to look for unusual sensory-related
processing. 'We just wanted to let the data talk to us. And that was
the main thing that popped out.'

This is an example of over-sensitivity in overdrive – and it helps
us to understand the importance of the neurotypical brain's ability
to step outside the immediate sensory world.

In theory, this finding could help to explain touch, sound and
visual over-responsivities. And the finding that levels of 'over-
connectivity' corresponded to the severity of autistic traits is
important. 'All of this speaks to this sense of invasion from the
outside world,' says Keysers, summing up. 'The external world has
a stronger pathway in their brain, and maybe they decouple their
internal processes from what's happening in the outside world less
routinely.' This fits, then, with the idea, mentioned in the last chapter,
that some autistic people develop alexithymia because they are overwhelmed by sensory signals of emotion.

In terms of other senses, touch over-sensitivities could have all kinds of impacts on a developing child.[35] Hating the feel of a nappy is one thing. But perceiving the stroke of your mother's hand as being like sandpaper is quite another, as the importance of human touch for typical social and emotional development is well-established. Strong over-sensitivities to lights and sounds could also impede the development of social skills because they make it hard to be in a public place, such as a shop or a school, where we learn patterns of interaction with others.

Caroline Robertson wants to get beyond intuitively appealing, broad-brush propositions like this, though, and drill down into the fine detail of sensory differences in people with autism. It's lab-based work like this, she thinks, that is shoring up the argument that for some people with autism, at least, sensory problems drive social difficulties.

There is now clear evidence, Robertson and Baron-Cohen think, that sensory processing differences in people who go on to be diagnosed with autism are observable early on (by around six months, before any social problems become apparent), and that the severity of these symptoms predicts how severe that individual's autism – including their level of social and also cognitive, thinking problems – will turn out to be.

Robertson herself is focusing on vision. Using a naturalistic 3D virtual world, she's finding that while non-autistic people tend to be drawn to faces and text, autistic people home in on regions of a scene that stand out because of their colour or their orientation, for example. This is a compelling demonstration, Robertson argues, of a 'detail-focused' visual preference in people with autism – the kind of focus that has led to a description of their being unable to 'see the wood for the trees'. (It's also evidence that, for them, faces are relatively less appealing; and if you grow up paying scant attention to faces, it could be tougher to learn to interpret the likely meaning of expressions or glances.)

Other work has found, meanwhile, that some people with autism have difficulty hearing the difference between speech sounds. That could easily make it harder not only to comprehend but also to form speech.

At a more fundamental level, there's evidence for altered levels of a key neurotransmitter called GABA in brain regions involved in processing sounds, visual signals and touch. It's work like this, write Caroline Robertson and Simon Baron-Cohen, that suggests the GABA system 'is key to the neurobiology of autism.'[36]

A so-called 'inhibitory' neurotransmitter, GABA 'calms' the nervous system. In theory, GABA dysfunction, causing an imbalance in neural excitation and inhibition, could explain sensory over- and under-responsivities. 'It could be that there are regional differences in this imbalance,' Robertson explains. 'For some people, somatosensation is really affected, and for others it's visual. It could also be that this imbalance leads to a vulnerability that is really sensitive to context – so for example, I'm not really sensitive to touch all the time, but when I'm being over-stimulated, I am more quick to reach the threshold of overload.'

Abnormal GABA levels have been linked to Attention Deficit Hyperactivity Disorder, too. This might explain why there are some shared traits – why sensory problems and attention difficulties are often present in both autism and ADHD, for example.[37]

Autism and ADHD are of course relatively common disorders. In the UK, there's evidence that 1.5 per cent of people fall on the autism spectrum, while an estimated one in 200 children under sixteen have been prescribed an ADHD drug.[38] A better understanding of the role of differences in sensing in explaining the classic symptoms of these disorders could, in theory, open up new routes to treatment. For example, targeting the way an 'over-sensitive' brain responds to sensory signals might reduce their intensity, making them less all-consuming, potentially helping people with a range of diagnoses.[39] This is something Elysa Marco is exploring in her lab right now.

More broadly, though, work like Elaine Aron's makes it clear that for all of us, how literally sensitive we are can have a profound influence on our lives. When I wrote in the introductory chapter about how our senses not only inform us but *form* us, it was this research, along with the wealth of studies finding differences between people concerning specific senses, that was foremost on my mind.

16

A Sense of Change

Now that these things have been defined, let us talk in general
about all perception . . . .

Aristotle, *De Anima*

Now that these things have been defined . . . let's take a look at
where our sense journey has brought us – and at the road that still
lies ahead.

First, it is clear we have many more than five senses. Aristotle's
model belongs in the history of science, not in our schools. As we've
learned about our senses, we've learned, too, that they have ancient
origins. Life forms that evolve together sense together, and we have
more in common, then, with everything from a bacterium to an
earthworm to a shrimp than we could have imagined.

Think back to those early life forms, which could detect changes
in their environment and in their own bodily state, and respond to
them – without the benefit of a brain. That history is written in
everything from the taste-like cells in our gut to smell receptors in
sperm. As we now know, conscious sensory perceptions reflect only
a fraction of what we detect. In fact, in terms of sensing, if you
remove central, conscious experiences from the picture, are we really
that radically different from a banana plant, or an oak?

One of the reasons plant-sensing has been so woefully under-
appreciated lies, unfortunately, at Aristotle's door, too: 'In plants we
do not find sensation nor any organ of sensation, nor any semblance
of it.'[1] This is an opinion that has persisted – though it's true that
not everyone has bought into it.

Take Prince Charles. He's long been ridiculed for talking to plants, because he believes it helps them to grow. (If you're anything like my age, perhaps you'll remember the *Spitting Image* satirical sketches of Charles, always talking to his plants.) If you do directly address a plant, you are of course bathing it in carbon dioxide. But everything from a *Mythbusters* episode to a recent study by the National Institute of Agricultural Biotechnology in South Korea has found that *recordings* of people talking, or of music, do encourage plant growth, perhaps by altering gene expression. The *Mythbusters* investigation found that plants grew better in greenhouses in which speech recordings were played, and grew better still when exposed to classical music – but death metal music was the most stimulating of all.[2]

As we know, plants can't hear, but they can sense vibrations, and so much more. And understanding how other life forms sense helps to put us in our place, don't you think? – As just another incredible, multi-sensing, utterly embedded respondent to sources of life-critical information on this, our planet Earth.

These ancient origins mean that of course sensing came a very long way before thinking. And this history means that information from our senses still feeds, all the time, into our decisions. In some circumstances, it makes our decisions for us. Whenever you go to fetch a glass of water, because you're feeling thirsty, or turn up the central heating against a chill, your senses are directing your actions. But as we now also appreciate, the influences of the senses on our minds can be so much more profound.

We understand now that sensory perceptions are vital for implicit learning – the sort that allows us to spot patterns in the world, without consciously knowing what they are. Yes, this style of learning might be relatively primitive, but we humans still use it all the time. And for people who are better at it, there are real advantages. Remember those London traders – their superior inner sensing drove their money-making 'smarts'. For all of us, though, this class of sensing is critical to our ability to feel emotions, and to empathise with other people. This is social sensing at its most extreme. Some researchers even think that it is signals from our bodies – from our organs and our muscles – that generate our perception of having a self – that create our subjective feeling of 'I'.[3]

What's also become apparent is that there is a huge spectrum of sensory experience, driven by variations in genes, life experience and culture. This knowledge should help us to be more tolerant of the opinions of our neighbours and friends. If one argues that a room is 'too warm', or that a meal is 'too salty' or that those flowers 'stink', and you don't feel the same way, perhaps you can resolve to agree to disagree – and even share the message that our perceptions of reality can be very different. Someone else's conflicting opinion shouldn't be taken, then, as an affront to your own good 'sense'. (Though you might have trouble convincing a top chef – even Jozef Youssef concedes, with a laugh, that he, like other chefs, thinks he knows precisely how much salt a dish needs to taste just right . . .)

However we sense the world, though, there are consequences for us that are hard to over-state. In some cases, they can be difficult to live with. Think of touch hyper-sensitivities in some children with Sensory Processing Disorder or autism, for example – or about how problems with processing sounds could contribute to psychosis. Other variations, though, can help to drive different kinds of experiences; Olympic athletes and Emmy-award-winning musicians will owe at least some of their success to the way they sense their inner and outer worlds. But even for those of us with milder differences, they can strike deep, profoundly influencing the way we interact with other people, and shaping our careers.

As we've learned, however, it's also eminently possible to alter the way that you sense – and so alter yourself. From Tim Birkhead clicking away in a university bathroom (he didn't want to tell me it was a bathroom – I guess he worried it might come across as unseemly to be experimenting in such a space – but it was) to exercising with your eyes shut, I've mentioned all kinds of methods to change your senses.

No, we can't all be prima ballerinas, or deep-sea divers, but we *can* train our vestibular senses and also proprioception, for body mapping, and also become more 'in touch' with our own hearts, and enjoy the physical and emotional benefits that this brings. We can learn to out-wit pain (at least, to some extent), and use our sense of temperature not just for practical purposes, but to influence our psychological state. We can improve, too, our sense of direction – and our ability to empathise with other people. And when it comes to fundamental shifts, do they get any more essential than literally learning to see the world differently?

With practice, we could also turn a very ordinary sense of smell into a thing of wonder. Nadjib Achaibou, the perfumer who would love us all to pick smell up from its gutter – and stop to sniff as we do so – tells me: 'I have this saying: you need to stop breathing and start smelling. You cannot stop breathing, you are always going to smell – so you might as well do it *consciously*.'

Of all the many ways to change our senses, there's one, though, that I haven't covered yet. In fact, I've been saving it. Because it's the fastest, and most potent of all.

$$*$$

If the doors of perception were cleansed every thing would appear to man as it is, Infinite. For man has closed himself up, till he sees all things thro' narrow chinks of his cavern.
William Blake, *The Marriage of Heaven and Hell*

When, in the spring of 1953, the writer Aldous Huxley took mescaline for the first time, it was as a willing, 'indeed eager' guinea pig for the British psychiatrist Humphry Osmond. Osmond believed that 'psychedelic' drugs (a term that he'd coined from the ancient Greek *psyche*, meaning mind or soul, and *deloun*, meaning show) could be useful for treating mental illness.[4] For his part, Huxley was keen to discover whether, through mescaline, he could break the bounds of his personal reality, to gain a profound insight into the mental life of a visionary, a medium or a mystic.

Huxley anticipated lying with his eyes shut, looking at wonderful visions of 'animated architectures' and 'symbolic dramas'. As he explains in *The Doors of Perception*, this didn't happen. But his altered perceptions of prosaic objects were revelatory. Huxley was captivated by the colours of flower petals and the drape of the cloth of his trousers. What he regarded, pre-mescaline, as a simple arrangement of a rose, a carnation and an iris in a glass vase was transformed: the carnation became 'a feathery incandescence', the iris 'smooth scrolls of sentient amethyst'.[5]

Countless people have written about their experiences on psyche-delics. But surely none as gorgeously as Huxley. He went on to experience a sense of unity with the physical world. He describes, for example, spending minutes – 'or several centuries' – not simply gazing at the bamboo legs of a chair but 'being them', his sense of self, and of a distinction between himself and those legs, having dissolved.

Humphry Osmond was interested in the potential power of psychedelics for understanding and treating mental illness. But when LSD and mescaline became associated with 1960s counter-culture, it became very difficult to obtain licences to administer them in experiments. Investigations into their psychiatric uses stopped. Recently, however, this has changed. More than six decades after *The Doors of Perception*, neuroscientific research is finally revealing the brain changes that underpin psychedelic drug experiences such as Huxley's.

In 2016, the first ever brain imaging studies of people on LSD were published. Consistent with similar studies using other classic hallucinogens (psilocybin from magic mushrooms and DMT from ayahuasca), the work shows that LSD creates greater disorder and also greater connectivity in the brain, linking regions that don't normally talk to each other.[6]

Such enhanced connectivity, or flexibility, has been associated with greater flexibility in thinking – with breaking free from formerly entrenched ideas. And precisely these impacts have been reported by people taking psychedelic drugs in the context of trials to inves-tigate their effects on depression and anxiety.

The LSD work revealed that the areas of the brain that become 'super-connected' include the insula, which of course receives sensory information and is important for emotion, and the frontoparietal cortex, which is associated with the representation of knowledge about the world, and also time. When the researchers looked at which brain networks became more engaged with these two regions under the influence of LSD, they identified four – and all involved sensory areas. In one, the sensorimotor cortex was important; two included regions of the visual cortex; in the fourth, the auditory cortex was a key node.

In 2019, another LSD study produced a fresh critical finding for this field. It showed that the drug initiates a kind of opening of the floodgates between the thalamus, which relays sensory information, and the cortex.[7] The resulting 'excessive' transmission of sensory information about the external and internal world could underpin not only the altered sensory perceptions characteristic of drugs such as LSD but also the sense of the dissolving of the ego – of becoming more at one with the physical world. The swelling of the sensory river to the cortex may, as Enzo Tagliazucchi at the Netherlands Institute for Neuroscience, lead author of the 2016 study, puts it, strengthen the link between our sense of self and the environment around us, 'potentially diluting the boundaries of our individuality'.

This is remarkable, isn't it? If Tagliazucchi is right, the volume of sensory information rushing through our brains informs our sense of separation – or otherwise – from the physical world around us.

There are other states in which people can report a state of transcendence. Exactly what happens during these experiences is not well understood. But they don't happen in hectic offices or pubs. They tend to take place in remote, natural settings – out in a national park on a perfect starlit night, perhaps. Or, from my own personal experience, at Uluru, in Australia's red heart. I'll never forget, more than twenty years ago now, standing at its base, looking up at slight clouds scudding through a pure blue sky, and feeling an overwhelmingly still sense of time having stopped, but also being infinite. 'I' felt myself to belong to something eternal. Psyche-shaking experiences like this are not instigated by our thinking brains. They're deeper. More 'primitive' . . . More sensory.

That sensation soon passed. And the pharmacological effects of psychedelic drugs of course wear off. But according to the testimony of numerous people who have taken part in recent psychotherapeutic drug trials, the psychological ramifications of intense ego-dissolution can be life-changing.[8] There are no end of case studies in support of this. For example, one participant in a psilocybin study run by a team at Imperial College London was a man who had been struggling with depression for thirty years. As he told the magazine *Mosaic*, he had virtually given up hope of overcoming it. But one dose of the drug changed everything: 'I couldn't believe how much it had changed so quickly. My approach to life, my attitude, my way of looking at the world, just everything, within a day.'[9]

Temporarily tamper with the way your brain receives sensory information, or the type of information that it receives, and you can, it seems, shock and re-form your entire outlook on life.

What this sort of tampering can't do, of course, is guide us towards an understanding of 'reality'. Huxley famously wrote that mescaline allowed him to see things 'as they really are'. We know that, clearly, this was not the case. Our senses can only give us imperfect, indirect contact with whatever is 'out there'. Psychedelics only replace one controlled hallucination of reality with another.

Still, Huxley's hopes for what mescaline might do for him will resonate with many of us. He had a volcanic desire to experience the world in different ways, to get around the 'reducing valve' of our own nervous systems and brains – to expand or even knock out new 'chinks' in our caverns.

Though we'll never know exactly what it's like to perceive infrared as a Western diamondback rattlesnake does, or sense the electrical field of a dandelion as a honeybee can, how wonderful it would be to have first-hand appreciation of other senses – to get a real hint of different ways of being, to borrow from Huxley, an animal on this particular planet.

What might come first, I wonder. Even if magnetoception is not something that we do naturally, once we understand just how other animals achieve it, it doesn't seem too sci-fi to assume that one day soon, humans could be tweaked to be capable of it, too. I also think of the mice that can now sense infrared. If that visual enhancement can work for them, in theory, it can work for us. And I remember that ancient animal, mentioned in Chapter 1, that first peeked above the water and glimpsed an entirely new world . . .

While research to get us to that stage edges ahead, we can surely still marvel at the true scope of our own existing sensory repertoire. Our new-found understanding of the senses we have is something that Aristotle – or even Sir Charles Scott Sherrington, for that matter – could never have imagined. We have gone from five human senses to thirty-two, and from a subordination of the senses to an appreciation of how they rule every aspect of our lives, profoundly influencing how we think, feel and behave.

There are, of course, still many questions. Some are focused on details and well-defined, such as: precisely which proteins constrain limb-location sensitivity? Others are more open, and more tantalising:

- To what extent does our individual taste receptor hardware influence our physical health?
- How will changing the way people sense their internal world affect their mental health?
- If we can prevent or fix deteriorations in sensing, what kind of a difference will that make to quality of life and brain health in older age?
- Even, do we have senses that are yet to be discovered?

With animals, there are some enormous known unknowns. For Tim Birkhead, the stand-out for birds is this: somehow, a flamingo wintering on the coast of southern Africa can sense when rain falls hundreds of kilometres away in shallow salt pans in Botswana and Namibia. However, it won't leave its wintering grounds unless *enough* rain has fallen to make the long flight inland worthwhile. How it detects when rain has fallen – and how much has come down – no one knows.

There is no equivalent known mystery for humans. But given the transformation in our sensory understanding in recent years, it's hard to resist wondering what kind of landmark discoveries are yet to be made.

In her 1889 article in *Science* in which she described work on 'an unknown organ of sense', Christine Ladd-Franklin wrote: 'In the frequent dwelling upon questions of development, which one cannot avoid in these days, one sometimes wonders whether the future is destined to endow man with any senses which he is not now in possession of.'

Perhaps, rather than being discovered, new human senses will be made. Still, if Ladd-Franklin were alive to ponder the same question today, the answer would be the same: indubitably, yes. And, from our modern standpoint, we can appreciate something else: that with extra senses will come not only new realities, but also new ways of being.

Who knows where in the future our senses will take us and what they will make us. But for now, I'm content to better understand my own extraordinary sensory world – to appreciate it, to revel in it, to use my knowledge of how it works, and also to do what I can to protect it.

And with that, I'm off to descend the dark stairs from my attic, check that my boys' bedrooms are warm enough, and if not, adjust the thermostats on their oil heaters, decide what I *really* need to eat for a late night snack, stretch my aching neck, then engage in my new resolution to stand on one leg while brushing my teeth – to do, in short, at least five every day, 'new' sense-dependent, eminently possible things before bed.

Acknowledgements

To all the researchers who took the time to explain their fascinating work to me – deepest thanks.

These chapters also benefit hugely from the personal stories. To Sue Barry, Nick Johnson, Nadjib Achaibou, Jozef Youssef, Stephen, Steph Singer, Yoko Ichino (via Lauren Godfrey), Herbert Nitsch, Fiona Torrance, Rachel Schneider and Lori Craven – thank you so much for sharing your experiences with me.

A few sections of this book have their origins in features that I wrote for *Mosaic*, the now sadly defunct magazine of the Wellcome Trust. Huge thanks to my wonderful editors there – Michael Regnier, Mun-Keat Looi and Chrissie Giles. Short segments of chapters 3 (Smell), 14 (Sensing Emotion) and 15 (Feeling Sensitive) first appeared in *Mosaic* features and are republished here under the Creative Commons Licence.

Research described in several stories that I wrote for the British Psychological Society's *Research Digest* also found its way into the book – thank you to my wonderful current and former BPS colleagues, Jon Sutton, Matt Warren and Christian Jarrett.

Kate Douglas, you are not only an excellent editor, but such a smart advisor and dear friend – thank you for all your commissions, all your counsel and all your support (and especially for regularly cycling over for lunch).

I am very fortunate to have other friends and family who were willing to share their expertise and give their time to check my writing. To Dr Jane Dixon, Dr Anu Carr, Dr Simon Carr and Dr Andrew Thorpe – enormous thanks. Thank you, too, Mr Bish.

To my dear friends and fellow science journalists and authors Gaia Vince and Jo Marchant, thank you for all your support from the very start (and only we know just how far back this book really goes . . .)

To Toby Mundy, my wonderful agent – thank you for your enthusiasm and support all the way through, and for finding what turned out to be the perfect home for this book – with John Murray. To Georgina Laycock, my publisher, sincere thanks for pushing the book into a different, much better form. And thank you to Abi Scruby for your detailed, enormously helpful editorial work in getting the book into shape.

Finally, eternal thanks to James, my husband and strongest supporter. And to Jakob and Lucas, my funny, smart, curious sons, for your love, and for expanding into every part of my life, even writing.

Notes

Author's note: Any quotes or comments about research in the text that are not referenced below are taken from interviews that I conducted personally.

Introduction

1 https://www.newscientist.com/article/dn17453-timeline-the-evolution-of-life/

2 Smith, C.U.M., *Biology of Sensory Systems*, 2nd edn, Wiley–Blackwell (2008).

3 Hug, Isabelle, et al., 'Second Messenger–Mediated Tactile Response by a Bacterial Rotary Motor', *Science* 358.6362 (2017): 531–4.

4 Haswell, Elizabeth S., Phillips, Rob and Rees, Douglas C., 'Mechanosensitive Channels: What Can They Do and How Do They Do It?', *Structure*, 19.10 (2011): 1,356–69.

5 Albert, D. J., 'What's on the Mind of a Jellyfish? A Review of Behavioural Observations on Aurelia sp. Jellyfish', *Neuroscience & Biobehavioral Reviews*, 35.3 (2011): 474–82.

6 Perbal, G. (2009), 'From ROOTS to GRAVI-1: Twenty-Five Years for Understanding How Plants Sense Gravity', *Microgravity Science and Technology*, 21.1–2 (2009): 3–10.

7 https://www.aao.org/eye-health/anatomy/rods

8 Howes, David (ed.), *The Varieties of Sensory Experience*, University of Toronto Press (1991); for a truly fascinating read, visit: http://www.sensorystudies.org/sensorial-investigations/doing-sensory-anthropology/

9 Chang, Yi-Shin, et al., 'Autism and Sensory Processing Disorders: Shared White Matter Disruption in Sensory Pathways but Divergent Connectivity in Social-Emotional Pathways', *PloS ONE*, 9.7 (2014): e103038.

Chapter 1: Sight

1 Schuergers, Nils, et al., 'Cyanobacteria Use Micro-Optics to Sense Light Direction.' *eLife* 5 (2016): e12620.

2 https://news.northwestern.edu/stories/2017/march/vision-not-limbs-led-fish-onto-land-385-million-years-ago/; MacIver, M. A., et al., 'Massive Increase in Visual Range Preceded the Origin of Terrestrial Vertebrates', *PNAS*, 11412 (2017): E2375–84.

3 Pearce, Eiluned, Stringer, Chris, and Dunbar, Robin I. M., 'New Insights Into Differences in Brain Organization Between Neanderthals and Anatomically Modern Humans', *Proceedings of the Royal Society B: Biological Sciences*, 280.1758 (2013), https://doi.org/10.1098/rspb.2013.0168; Pearce, Eiluned, and Dunbar, Robin, 'Latitudinal Variation in Light Levels Drives Human Visual System Size', *Biology Letters*, 8.1 (2012): 90–3.

4 Caval-Holme, Franklin, and Feller, Marla B., 'Gap Junction Coupling Shapes the Encoding of Light in the Developing Retina', *Current Biology*, 29.23 (2019): 4,024–35.

5 Hyvärinen, Lea, et al., 'Current Understanding of What Infants See', *Current Ophthalmology Reports*, 2.4 (2014): 142–9.

6 Douglas, R. H., and Jeffery, G., 'The Spectral Transmission of Ocular Media Suggests Ultraviolet Sensitivity is Widespread Among Mammals', *Proceedings of the Royal Society B: Biological Sciences*, 281.1780 (2014) https://doi.org/10.1098/rspb.2013.2995.

7 https://www.newscientist.com/article/mg22630170-400-eye-of-the-beholder-how-colour-vision-made-us-human/#ixzz6CVtbVn6x

8 https://ghr.nlm.nih.gov/condition/color-vision-deficiency#statistics

9 Osnos, Evan, 'Can Mark Zuckerberg Fix Facebook Before It Breaks Democracy?', *New Yorker*, 10 September 2018.

10 Hunt, David M., et al., 'The Chemistry of John Dalton's Color Blindness', *Science* 267.5200 (1995): 984–8.

11 Jordan, Gabriele, et al., 'The Dimensionality of Color Vision in Carriers of Anomalous Trichromacy', *Journal of Vision*, 10.8 (2010): 12–12.

12 Winderickx, Joris, et al., 'Polymorphism in Red Photopigment Underlies Variation in Colour Matching', *Nature*, 356.6368 (1992): 431–3.

13 Provencio, Ignacio, et al., 'Melanopsin: An Opsin in Melanophores, Brain, and Eye', *Proceedings of the National Academy of Sciences*, 95.1 (1998): 340–5.

14　Roecklein, Kathryn A., et al., 'A Missense Variant (P10L) of the Melanopsin (OPN4) Gene in Seasonal Affective Disorder', *Journal of Affective Disorders*, 114.1–3 (2009): 279–85.

15　Terman, Michael, and McMahan, Ian, *Chronotherapy*, Penguin (2012).

16　Sherman, S., and Guillery, R., 'The Role of the Thalamus in the Flow of Information to the Cortex', *Philosophical Transactions of the Royal Society B: Biological Sciences*, 357.1428 (2002): 1,695–1,708, https://doi.org/10.1098/rstb.2002.1161

17　Huff, T., Mahabadi, N., and Tadi, P., 'Neuroanatomy, Visual Cortex', *StatPearls* (2019), pmid: 29494110.

18　Cicmil, Nela, and Krug, Kristine, 'Playing the Electric Light Orchestra: How Electrical Stimulation of Visual Cortex Elucidates the Neural Basis of Perception', *Philosophical Transactions of the Royal Society B: Biological Sciences*, 370.1677 (2015), https://doi.org/10.1098/rstb.2014.0206

19　Kanwisher, N., Stanley, D., and Harris, A., 'The Fusiform Face area is Selective for Faces Not Animals', *Neuroreport*, 10.1 (1999): 183–7.

20　Cuaya, L. V., Hernández-Pérez, R., and Concha, L., 'Our Faces in the Dog's Brain: Functional Imaging Reveals Temporal Cortex Activation During Perception of Human Faces', *PloS ONE*, 11.3 (2016): e0149431.

21　McCrae, Robert R., 'Creativity, Divergent Thinking, and Openness to Experience', *Journal of Personality and Social Psychology*, 52.6 (1987): 1,258–68.

22　Antinori, Anna, Carter, Olivia L., and Smillie, Luke D., 'Seeing It Both Ways: Openness to Experience and Binocular Rivalry Suppression', *Journal of Research in Personality*, 68 (2017): 15–22.

23　Davidoff, Jules, Davies, Ian, and Roberson, Debi, 'Colour Categories in a Stone-Age Tribe', *Nature*, 398.6724 (1999): 203–4.

24　Goldstein, Julie, Davidoff, Jules, and Roberson, Debi, 'Knowing Color Terms Enhances Recognition: Further Evidence From English and Himba', *Journal of Experimental Child Psychology*, 102.2 (2009): 219–38.

25　https://digest.bps.org.uk/2018/11/02/your-native-language-affects-what-you-can-and-cant-see/

26　https://www.urmc.rochester.edu/del-monte-neuroscience/neuroscience-blog/december-2018/the-science-of-seeing-art-and-color.aspx

27　http://persci.mit.edu/gallery/checkershadow

28　https://www.ted.com/talks/anil_seth_how_your_brain_hallucinates_

your_conscious_reality/footnotes?fbclid=IwAR1F_kZNByH-hPf-7vRI9aTuW2nzbsBKITZIRBmgGFS8hMo2MNcrQGOUUgw

29   Otten, Marte, et al., 'The Uniformity Illusion: Central Stimuli Can Determine Peripheral Perception', *Psychological Science*, 28.1 (2017): 56–68.

30   https://jov.arvojournals.org/SS/thedress.aspx

31   The phrase 'controlled hallucination' also appears in a research paper by Rick Grush (http://escholarship.org/uc/item/15t2595z) who in turn attributes the term to a talk at the University of California San Diego given by Ramesh Jain – which was never recorded – and there the trail fades out.

32   http://www.sussex.ac.uk/synaesthesia/faq#howcommon

33   Simner, Julia, and Logie, Robert H., 'Synaesthetic Consistency Spans Decades in a Lexical–Gustatory Synaesthete', *Neurocase*, 13.5–6 (2008): 358–65.

34   Simner, Julia, et al., 'Synaesthesia: The Prevalence of Atypical Cross-Modal Experiences', *Perception*, 35.8 (2006): 1,024–33.

35   Bosley, Hannah G., and Eagleman, David M., 'Synesthesia in Twins: Incomplete Concordance in Monozygotes Suggests Extragenic Factors', *Behavioural Brain Research*, 286 (2015): 93–6.

36   Simner, Julia, et al., 'Early Detection of Markers for Synaesthesia in Childhood Populations', *Brain*, 132.1 (2009): 57–64; Simner, Julia, and Bain, Angela E., 'A Longitudinal Study of Grapheme-Color Synesthesia in Childhood: 6/7 Years to 10/11 Years', *Frontiers in Human Neuroscience*, 7 (2013), https://doi.org/10.3389/fnhum.2013.00603

37   Farina, Francesca R., Mitchell, Kevin J., and Roche, Richard A. P., 'Synaesthesia Lost and Found: Two Cases of Person-and Music-Colour Synaesthesia', *European Journal of Neuroscience*, 45.3 (2017): 472–7.

38   Ward, Jamie, et al., 'Atypical Sensory Sensitivity as a Shared Feature between Synaesthesia and Autism', *Scientific Reports*, 7 (2017), https://doi.org/10.1038/srep41155

39   Tilot, A. K., et al., 'Rare Variants in Axonogenesis Genes Connect Three Families with Sound-Color Synesthesia', *Proceedings of the National Academy of Sciences*, 115.12 (2018): 3,168–73.

40   Shriki, Oren, Sadeh, Yaniv, and Ward, Jamie, 'The Emergence of Synaesthesia in a Neuronal Network Model Via Changes in Perceptual Sensitivity and Plasticity', *PLoS Computational Biology*, 12.7 (2016), https://doi.org/10.1371/journal.pcbi.1004959

41   Forest, Tess Allegra, et al., 'Superior Learning in Synesthetes: Consistent

Grapheme-Color Associations Facilitate Statistical Learning', *Cognition*, 186 (2019): 72–81.

42 Treffert, Darold A., 'The Savant Syndrome: An Extraordinary Condition. A Synopsis: Past, Present, Future', *Philosophical Transactions of the Royal Society B: Biological Sciences*, 364.1522 (2009): 1,351–7.

43 Baron-Cohen, Simon, et al., 'Savant Memory in a Man with Colour Form-Number Synaesthesia and Asperger', *Journal of Consciousness Studies*, 14.9–10 (2007): 237–51.

44 Baron-Cohen, Simon, et al., 'Is Synaesthesia More common in Autism?', *Molecular Autism*, 4.1 (2013): 40; Hughes, James E. A., et al., 'Is Synaesthesia More Prevalent in Autism Spectrum Conditions? Only Where There is Prodigious Talent', *Multisensory Research*, 30.3–5 (2017): 391–408.

45 Gomez, J., Barnett, M., and Grill-Spector, K., 'Extensive Childhood Experience with Pokémon Suggests Eccentricity Drives Organization of Visual Cortex', *Nature Human Behaviour*, 3.6 (2019): 611–24.

46 http://www.oepf.org/sites/default/files/journals/jbo-volume-14-issue-2/14-2%20Godnig.pdf

47 https://nei.nih.gov/news/briefs/defective_lens_protein

48 Patel, Ilesh, and West, Sheila K., 'Presbyopia: Prevalence, Impact, and Interventions', *Community Eye Health*, 20.63 (2007): 40.

49 Zhou, Zhongqiang, et al., 'Pilot Study of a Novel Classroom Designed to Prevent Myopia by Increasing Children's Exposure to Outdoor Light', *PLoS ONE*, 12.7 (2017): e0181772.

50 Williams, Katie M., et al., 'Increasing Prevalence of Myopia in Europe and the Impact of Education', *Ophthalmology*, 122.7 (2015): 1,489–97.

51 See, for example, Dolgin, Elie, 'The Myopia Boom', *Nature*, 519.7543 (19 March 2015): 276–8, https://doi.org/10.1038/519276a

52 Wu, Pei-Chang, et al., 'Outdoor Activity During Class Recess Reduces Myopia Onset and Progression in School Children', *Ophthalmology*, 120.5 (2013): 1,080–5.

53 Williams, Paul T., 'Walking and Running are Associated with Similar Reductions in Cataract Risk', *Medicine and Science in Sports and Exercise*, 45.6 (2013): 1,089.

54 Smith, Annabelle, K., 'A WWII Propaganda Campaign Popularized the Myth That Carrots Help You See in the Dark', *Smithsonian Magazine*, 13 August 2013.

55 Harrison, Rhys, et al., 'Blindness Caused by a Junk Food Diet', *Annals of Internal Medicine*, 171.11 (2019): 859–61.

56 Gislén, Anna, et al., 'Superior Underwater Vision in a Human Population of Sea Gypsies', *Current Biology*, 13.10 (2003): 833–6.

57 Gislén, Anna, et al., 'Visual Training Improves Underwater Vision in Children', *Vision Research*, 46.20 (2006): 3,443–50.

58 Sacks, O., 'Stereo Sue', New Yorker, 12 June 2006; see also: Barry S. R., *Fixing My Gaze: A Scientist's Journey into Seeing in Three Dimensions*, Basic Books (2009).

59 Barry, Susan R., and Bridgeman, Bruce, 'An Assessment of Stereovision Acquired in Adulthood', *Optometry and Vision Science*, 94.10 (2017): 993–9.

60 Camacho-Morales, Rocio, et al., 'Nonlinear Generation of Vector Beams From AlGaAs Nanoantennas', *Nano Letters*, 16.11 (2016): 7,191–7.

61 Ma, Yuqian, et al., 'Mammalian Near-Infrared Image Vision Through Injectable and Self-Powered Retinal Nanoantennae', *Cell*, 177.2 (2019): 243–55.

62 https://www.eurekalert.org/pub_releases/2019-08/acs-ncs071819.php

63 Gu, Leilei, et al., 'A Biomimetic Eye With a Hemispherical Perovskite Nanowire Array Retina', *Nature*, 581 (2020): 278–82.

64 Seth, Anil K., 'From Unconscious Inference to the Beholder's Share: Predictive Perception and Human Experience', *European Review*, 27.3 (2019): 378–410.

Chapter 2: Hearing

1 See https://www.calacademy.org/explore-science/do-plants-hear; also Jung, Jihye, et al., 'Beyond Chemical Triggers: Evidence for Sound-Evoked Physiological Reactions in Plants', *Frontiers in Plant Science*, 9 (2018), https://doi.org/10.3389/fpls.2018.00025

2 Appel, H. M., and Cocroft, R. B., 'Plants Respond to Leaf Vibrations Caused by Insect Herbivore Chewing', *Oecologia* (2014), https://doi.org/10.1007/s00442-014-2995-6

3 https://evolution.berkeley.edu/evolibrary/article/evograms_05

4 http://www.shark.ch/Information/Senses/index.html

5 https://www.phon.ucl.ac.uk/courses/spsci/acoustics/week2-9.pdf

6 DeCasper, Anthony J., and Fifer, William P., 'Of Human Bonding: Newborns Prefer Their Mothers' Voices', *Science*, 208.4448 (1980): 1,174–6; and, on the importance of that study, Busnel, Marie-Claire,

et al., 'Tony DeCasper, the Man Who Changed Contemporary Views on Human Fetal Cognitive Abilities', *Developmental Psychobiology*, 59.1 (2017): 135–9.

7 Heinonen-Guzejev, Marja, et al., 'Genetic Component of Noise Sensitivity', *Twin Research and Human Genetics*, 8.3 (2005): 245–9.

8 https://digest.bps.org.uk/2019/10/04/harsh-sounds-like-screams-hijack-brain-areas-involved-in-pain-and-aversion-making-them-impossible-to-ignore/

9 See https://www.psychologytoday.com/gb/blog/music-matters/201407/do-chimpanzees-music

10 Norman-Haignere, Sam V., et al., 'Divergence in the Functional Organization of Human and Macaque Auditory Cortex Revealed by fMRI Responses to Harmonic Tones', *Nature Neuroscience*, 22.7 (2019): 1,057–60; https://www.sciencedaily.com/releases/2019/07/190711111913.htm

11 See https://digest.bps.org.uk/2019/10/17/culture-plays-an-important-role-in-our-perception-of-musical-pitch-according-to-study-of-boliv-ias-tsimane-people/ ; Jacoby, N., et al., 'Universal and Non-Universal Features of Musical Pitch Perception Revealed by Singing', *Current Biology*, 29.19 (2019): 3,229–43.e12.

12 https://noobnotes.net/dancing-queen-abba/

13 Dolscheid, S., et al., 'The Thickness of Musical Pitch: Psychophysical Evidence for Linguistic Relativity', *Psychological Science*, 24.5 (2013): 613–21.

14 Dolscheid, S., et al., 'Prelinguistic Infants Are Sensitive to Space-Pitch Associations Found Across Cultures', *Psychological Science*, 25.6 (2014): 1,256–61.

15 Tajadura-Jiménez, Ana, et al., 'As Light as Your Footsteps: Altering Walking Sounds to Change Perceived Body Weight, Emotional State and Gait', *Proceedings of the 33rd Annual ACM Conference on Human Factors in Computing Systems*, Association for Computing Machinery (2015).

16 Powers, Albert R., Mathys, Christoph, and Corlett, P. R., 'Pavlovian Conditioning-Induced Hallucinations Result from Overweighting of Perceptual Priors', *Science*, 357.6351 (2017): 596–600.

17 Woods, Angela, et al., 'Experiences of Hearing Voices: Analysis of a Novel Phenomenological Survey', *Lancet Psychiatry*, 2.4 (2015): 323–31.

18 See McCarthy-Jones, Simon, et al., 'A new Phenomenological Survey of Auditory Hallucinations: Evidence for Subtypes and Implications for Theory and Practice', *Schizophrenia Bulletin*, 40.1 (2014): 231–5.

19 Ford, J. M., and Mathalon, D. H., 'Anticipating the Future: Automatic Prediction Failures in Schizophrenia', *International Journal of Psychophysiology*, 83.2 (2012): 232–9.

20 Sterzer, Philipp, et al., 'The Predictive Coding Account of Psychosis', *Biological Psychiatry*, 84.9 (2018): 634–43; Frith, Chris, *Making Up the Mind*, 1st edn, Blackwell Publishing (2007); Corlett, Philip R., et al., 'Hallucinations and Strong Priors', *Trends in Cognitive Sciences*, 23.2 (2019): 114–27.

21 Marshall, Amanda C., Gentsch, Antje, and Schütz-Bosbach, Simone, 'The Interaction Between Interoceptive and Action States Within a Framework of Predictive Coding', *Frontiers in Psychology*, 9 (2018): 180.

22 Klaver, M., and Dijkerman, H. C., 'Bodily Experience in Schizophrenia: Factors Underlying a Disturbed Sense of Body Ownership', *Frontiers in Human Neuroscience*, 10 (2016): 305.

23 Andrade, G. N., et al., 'Atypical Visual and Somatosensory Adaptation in Schizophrenia-Spectrum Disorders', *Translational Psychiatry*, 6 (2016): e804, https://doi.org/10.1038/tp.2016.63

24 Hanumantha, K., Pradhan, P. V., and Suvarna, B., 'Delusional Parasitosis – Study of 3 Cases', *Journal of Postgraduate Medicine*, 40.4 (1994): 222.

25 Ross, L. A., et al., 'Impaired Multisensory Processing in Schizophrenia: Deficits in the Visual Enhancement of Speech Comprehension Under Noisy Environmental Conditions', *Schizophrenia Research*, 97.1–3 (2007): 173–83.

26 Leitman, David I., et al., 'Sensory Contributions to Impaired Prosodic Processing in Schizophrenia', *Biological Psychiatry*, 58.1 (2005): 56–61.

27 https://www.birmingham.ac.uk/Documents/college-social-sciences/education/victar/thomas-pocklington-20-case-studies.pdf

28 Huber, Elizabeth, et al., 'Early Blindness Shapes Cortical Representations of Auditory Frequency Within Auditory Cortex', *Journal of Neuroscience*, 39.26 (2019): 5,143–52.

29 Stephan, Yannick, et al., 'Sensory Functioning and Personality Development Among Older Adults', *Psychology and Aging*, 32.2 (2017): 139–147.

30 https://www.nidcd.nih.gov/health/hearing-loss-older-adults

31 Lin, F. R., et al., 'Hearing Loss and Incident Dementia', *Archives of Neurology*, 68.2 (2011): 214–20.

32 https://news.osu.edu/subtle-hearing-loss-while-young-changes-brain-function-study-finds/

33 See https://digest.bps.org.uk/2020/05/27/gradual-hearing-loss-reorganises-brains-sensory-areas-and-impairs-memory-in-mice/;
    Beckmann, D., et al., 'Hippocampal Synaptic Plasticity, Spatial Memory, and Neurotransmitter Receptor Expression Are Profoundly Altered by Gradual Loss of Hearing Ability', *Cerebral Cortex*, 30.8 (2020): 4,581–96.

34 Huber, Elizabeth, et al., 'Early Blindness Shapes Cortical Representations of Auditory Frequency Within Auditory Cortex', *Journal of Neuroscience*, 39.26 (2019): 5,143–52.

35 See Walsh, R. M., et al., 'Bomb Blast Injuries to the Ear: The London Bridge Incident Series', *Emergency Medicine Journal*, 12.3 (1995): 194–8.

36 http://www.euro.who.int/en/health-topics/environment-and-health/noise; see also https://www.nidcd.nih.gov/health/noise-induced-hearing-loss; https://www.who.int/mediacentre/news/releases/2015/ear-care/en/

37 http://www.uzh.ch/orl/dga2006/programm/wissprog/Fleischer.pdf; https://www.newscientist.com/article/mg18224492-300-bang-goes-your-hearing-if-you-dont-exercise-your-ears/

38 Fredriksson, S., Kim, et al., 'Working in Preschool Increases the Risk of Hearing-Related Symptoms: A Cohort Study Among Swedish Women', *International Archives of Occupational and Environmental Health*, 92.8: (2019): 1,179–90.

39 See, for example, https://www.newyorker.com/magazine/2019/05/13/is-noise-pollution-the-next-big-public-health-crisis

40 Curhan, Sharon G., et al., 'Body Mass Index, Waist Circumference, Physical Activity, and Risk of Hearing Loss in Women', *American Journal of Medicine*, 126.12 (2013), https://doi.org/10.1016/j.amjmed.2013.04.026

41 Curhan, Sharon G., et al., 'Adherence to Healthful Dietary Patterns is Associated With Lower Risk of Hearing Loss in Women', *Journal of Nutrition*, 148.6 (2018): 944–51.

42 Anderson, Samira, et al., 'Reversal of Age-Related Neural Timing Delays With Training', *Proceedings of the National Academy of Sciences*, 110.11 (2013): 4,357–62; Song, Judy H., et al., 'Plasticity in the Adult Human Auditory Brainstem Following Short-Term Linguistic Training', *Journal of Cognitive Neuroscience*, 20.10 (2008): 1,892–902.

43 https://www.youtube.com/watch?v=lAtVOKo4XvA; see also Kish's Ted talk, https://www.youtube.com/watch?v=uHoaihGWB8U

44 See https://www.dur.ac.uk/research/news/item/?itemno=34855

45 Birkhead, T., *Bird Sense: What It's Like to Be a Bird*, Bloomsbury (2013).

Chapter 3: Smell

1   https://www.facebook.com/sheriffcitrus/posts/do-you-have-a-scent-preservation-kitk9-ally-hopes-that-you-dolast-night-k9-ally-/1443416362380828/

2   Porter, Jess, et al., 'Mechanisms of Scent-Tracking in Humans', *Nature Neuroscience*, 10.1 (2007): 27–9; 'People Track Scents in the Same Way as Dogs', https://www.nature.com/news/2006/061211/full/061211-18.html

3   Louden, Robert B., ed., *Kant: Anthropology from a Pragmatic Point of View*, Cambridge Texts in the History of Philosophy, Cambridge University Press (2006).

4   McGann, John P., 'Poor Human Olfaction is a 19th-Century Myth', *Science*, 356.6338 (2017), https://doi.org/10.1126/science.aam7263

5   'The Olfactory Epithelium and Olfactory Receptor Neurons', *Neuroscience*, 2nd edn, Sinauer Associates (2001).

6   Reindert Nijland, and Burgess, Grant, 'Bacterial Olfaction', *Biotechnology Journal*, 5.9 (2010): 974–977.

7   Nagayama, S., Homma, R., and Imamura, F., 'Neuronal Organization of Olfactory Bulb Circuits', *Frontiers in Neural Circuits*, 8.98 (2014), https://doi.org/10.3389/fncir.2014.00098

8   Li, Wen, et al., 'Right Orbitofrontal Cortex Mediates Conscious Olfactory Perception', *Psychological Science*, 21.10 (2010): 1,454–63.

9   Bushdid, C., et al., 'Humans Can Discriminate More Than 1 Trillion Olfactory Stimuli', *Science*, 343.6177 (2014): 1,370–2.

10 Hoover, Kara C., et al., 'Global Survey of Variation in a Human Olfactory Receptor Gene Reveals Signatures of Non-Neutral Evolution', *Chemical Senses*, 40.7 (2015): 481–8.

11 'Evolution of Primate Sense of Smell and Full Trichromatic Color Vision', *PLoS Biology*, 2.1 (2004): e33; https://doi.org/10.1371/journal.pbio.0020033

12 Hughes, Graham M., Teeling, Emma C., and Higgins, Desmond G., 'Loss of Olfactory Receptor Function in Hominin Evolution', *PloS ONE*, 9.1 (2014): e84714.

13 Lee, David S., Kim, Eunjung, and Schwarz, Norbert, 'Something Smells Fishy: Olfactory Suspicion Cues Improve Performance on the Moses Illusion and Wason Rule Discovery Task', *Journal of Experimental Social Psychology*, 59 (2015): 47–50.

14 See, for example, Schwarz, Norbert, et al., 'The Smell of Suspicion: How the Nose Curbs Gullibility', *The Social Psychology of Gullibility: Fake News, Conspiracy Theories, and Irrational Beliefs*, Routledge (2019): 234–52.

15 Mainland, Joel D., et al., 'The Missense of Smell: Functional Variability in the Human Odorant Receptor Repertoire', *Nature Neuroscience*, 17.1 (2014): 114–20.

16 Wedekind, Claus, et al., 'MHC-Dependent Mate Preferences in Humans', *Proceedings of the Royal Society B: Biological Sciences*, 260.1359 (1995): 245–9.

17 Keller, Andreas, et al., 'Genetic Variation in a Human Odorant Receptor Alters Odour Perception', Nature, 449.7161 (2007): 468–72; and related *Nature* news story, https://www.nature.com/news/2007/070910/full/070910-15.html

18 See Spinney, L., 'You Smell Flowers, I Smell Stale Urine', *Scientific American*, 1 February 2011.

19 https://embryology.med.unsw.edu.au/embryology/index.php/Sensory_-_Smell_Development

20 Lipchock, Sarah V., Reed, Danielle R., and Mennella, Julie A., 'The Gustatory and Olfactory Systems During Infancy: Implications for Development of Feeding Behaviors in the High-Risk Neonate', *Clinics in Perinatology*, 38.4 (2011): 627–41.

21 Mennella, J. A., Jagnow, C. P. and Beauchamp, G. K., 'Prenatal and Postnatal Flavor Learning by Human Infants', *Pediatrics*, 107.6 (2001): e88, https://doi.org/10.1542/peds.107.6.e88

22 Majid, Asifa and Burenhult, Niclas, 'Odors are Expressible in Language, as Long as You Speak the Right Language', *Cognition*, 130.2 (2014): 266–70.

23 Majid, Asifa, et al., 'Olfactory Language and Abstraction Across Cultures', *Philosophical Transactions of the Royal Society B: Biological Sciences*, 373.1752 (2018): https://doi.org/10.1098/rstb.2017.0139

24 Majid, Asifa, and Krupse, Nicole, 'Hunter-Gatherer Olfaction is Special', *Current Biology*, 28.3 (2018): 409–413, https://doi.org/10.1016/j.cub.2017.12.014

25 Hippocratic Corpus, Prognosticon, cited in Bradley, Mark, ed., *Smell and the Ancient Senses*, Routledge (2014).

26 See Bradley, Mark, ed., *Smell and the Ancient Senses*, Routledge (2014).

27 See, for example, Willis, Carolyn M., et al., 'Volatile Organic Compounds as Biomarkers of Bladder Cancer: Sensitivity and Specificity Using Trained Sniffer dogs', *Cancer Biomarkers*, 8.3 (2011): 145–53.

28 https://www.parkinsons.org.uk/news/meet-woman-who-can-smell-parkinsons

29 Trivedi, Drupad K., et al., 'Discovery of Volatile Biomarkers of Parkinson's Disease From Sebum', *ACS Central Science*, 5.4 (2019): 599–606.

30 Beauchamp, G., *Odor Signals of Immune Activation and CNS Inflammation*, Monell Chemical Senses Center Philadelphia, P.A. (2014).

31 Ferdenzi, Camille, Licon, Carmen, and Bensafi, Moustafa, 'Detection of Sickness in Conspecifics Using Olfactory and Visual Cues', *Proceedings of the National Academy of Sciences*, 114.24 (2017): 6,157–9.

32 Gervasi, S. S., et al., 'Sharing an Environment With Sick Conspecifics Alters Odors of Healthy Animals', *Scientific Reports*, 8.1 (2018): 1–13.

33 Szawarski, Piotr, 'Classic Cases Revisited: Oscar the Cat and Predicting Death', *Journal of the Intensive Care Society*, 17.4 (2016): 341–5.

34 Parmentier, M., Libert, F., et al., 'Expression of Members of the Putative Olfactory Receptor Gene Family in Mammalian Germ Cells', *Nature*, 355.6359 (1992): 453–5.

35 Vanderhaeghen, P., et al, 'Specific Repertoire of Olfactory Receptor Genes in the Male Germ Cells of Several Mammalian Species', *Genomics*, 39.3 (1997): 239–46.

36 Pluznick, Jennifer L., 'Renal and Cardiovascular Sensory Receptors and Blood Pressure Regulation', *American Journal of Physiology-Renal Physiology*, 305.4 (2013): F439–44.

37 Abaffy, Tatjana, 'Human Olfactory Receptors Expression and Their Role in Non-Olfactory Tissues: A Mini-Review', *Journal of Pharmacogenomics & Pharmacoproteomics*, 6.4 (2015): 1.

38 Zapiec, Bolek, et al., 'A Ventral Glomerular Deficit in Parkinson's Disease Revealed by Whole Olfactory Bulb Reconstruction', *Brain*, 140.10 (2017): 2,722–36.

39 See, for example, https://www.monellfoundation.org/index.php/the-monell-anosmia-project/

40 See, for example, Seow, Yi-Xin, Ong, Peter K. C., and Huang, Dejian, 'Odor-Specific Loss of Smell Sensitivity With Age as Revealed by the Specific Sensitivity Test', *Chemical Senses*, 41.6 (2016): 487–95.

41 Ibid.

42 Lecuyer Giguère, Fanny, et al., 'Olfactory, Cognitive and Affective Dysfunction Assessed 24 Hours and One Year After a Mild Traumatic Brain Injury (mTBI)', *Brain Injury*, 33.9 (2019): 1,184–93.

43 Martial's description of Thais is from Bradley, Mark, ed., *Smell and the Ancient Senses* (2014).

44 Corbin, A,, *The Foul and the Fragrant*, Harvard University Press (1986).

45 Just like any grumpy old man, Pliny even complained that the perfumes favoured in Rome were made from ingredients from far-flung countries: 'Not one bit of this perfume is produced in Italy, the conqueror of the world, not even within the confines of Europe.' And I can't omit this wonderful protest: 'By Hercules, people even add perfumes to their drinks!'

46 http://www.sirc.org/publik/smell.pdf

47 Liu, Bojing, et al., 'Relationship Between Poor Olfaction and Mortality Among Community-Dwelling Older Adults: A Cohort Study', *Annals of Internal Medicine*, 170.10 (2019): 673–81.

48 Holbrook, Eric H., et al., 'Induction of Smell Through Transethmoid Electrical Stimulation of the Olfactory Bulb', *International Forum of Allergy & Rhinology*, 9.2 (2019): 158–64.

49 Bendas, J., Hummel, T., & Croy, I., 'Olfactory Function Relates to Sexual Experience in Adults', *Archives of Sexual Behavior*, 47.5 (2018): 1333–9.

50 See http://centreforsensorystudies.org/occasional-papers/sensing-cultures-cinema-ethnography-and-the-senses/

51 Majid, Asifa, and Levinson, Stephen C., 'The Senses in Language and Culture', *The Senses and Society*, 6.1 (2011): 5–18.

Chapter 4: Taste

1 Spence, Charles, 'Oral Referral: On the Mislocalization of Odours to the Mouth', *Food Quality and Preference*, 50 (2016): 117–28.

2 Breslin, Paul A. S., 'An Evolutionary Perspective on Food and Human Taste', *Current Biology*, 23.9 (2013): R409–18.

3 Keast, Russell S. J., and Costanzo, Andrew, 'Is Fat the Sixth Taste Primary? Evidence and Implications', *Flavour*, 4.1 (2015): 5; Besnard, Philippe, Passilly-Degrace, Patricia, and Khan, Naim A., 'Taste of Fat: A Sixth Taste Modality?', *Physiological Reviews*, 96.1 (2016): 151–76.

Asifa Majid and Stephen Levinson have analysed cross-linguistic data (taken from widely different cultural contexts) on tastes, and, interestingly, this work supports the idea that sweet, salt, sour and bitter are basic tastes, with umami and fatty 'likely' basic tastes as well.

4 https://www.monell.org/news/fact_sheets/monell_taste_primer

5 Ibid.

6 Mainland, Joel D., and Matsunami, Hiroaki, 'Taste Perception: How Sweet It Is (To Be Transcribed by You)', *Current Biology*, 19.15 (2009): R655–6.

7 Lindemann, Bernd, Ogiwara, Yoko, and Ninomiya, Yuzo, 'The Discovery of Umami', *Chemical Senses*, 27.9 (2002): 843–4.

8 https://www.sciencedirect.com/topics/neuroscience/tas1r1

9 Chandrashekar, Jayaram, et al., 'The Cells and Peripheral Representation of Sodium Taste in Mice', *Nature*, 464.7286 (2010): 297–301; Lewandowski, Brian C., et al., 'Amiloride-Insensitive Salt Taste is Mediated by Two Populations of Type III Taste Cells With Distinct Transduction Mechanisms', *Journal of Neuroscience*, 36.6 (2016): 1,942–53.

10 Huang, Angela L., et al., 'The Cells and Logic for Mammalian Sour Taste Detection', *Nature*, 442.7105 (2006): 934–8; Challis, Rosemary C., and Ma, Minghong, 'Sour Taste Finds Closure in a Potassium Channel', *Proceedings of the National Academy of Sciences*, 113.2 (2016): 246–7.

11 Jaggupilli, A., et al., 'Bitter Taste Receptors: Novel Insights into the Biochemistry and Pharmacology', *International Journal of Biochemistry & Cell Biology*, 77 (2016): 184–96.

12 See Sagioglou, Christina, and Greitemeyer, Tobias, 'Individual Differences in Bitter Taste Preferences are Associated With Antisocial Personality Traits', *Appetite*, 96 (2016): 299–308.

13 Lachenmeier, Dirk W., 'Wormwood (Artemisia absinthium L.) – A Curious Plant With Both Neurotoxic and Neuroprotective Properties?', *Journal of Ethnopharmacology*, 131.1 (2010): 224–27.

14 Laffitte, Anni, Neiers, Fabrice, and Briand, Loïc, 'Functional Roles of the Sweet Taste Receptor in Oral and Extraoral Tissues', *Current Opinion in Clinical Nutrition and Metabolic Care*, 17.4 (2014): 379.

15 Benford, H., et al., 'A Sweet Taste Receptor-Dependent Mechanism of Glucosensing in Hypothalamic Tanycytes', *Glia*, 65.5 (2017): 773–89.

16 Lazutkaite, G., et al., 'Amino Acid Sensing in Hypothalamic Tanycytes Via Umami Taste Receptors', *Molecular Metabolism*, 6.11 (2017): 1,480–92.

17 Kotrschal, K., 'Ecomorphology of Solitary Chemosensory Cell Systems in Fish: A Review', in *Ecomorphology of Fishes*, ed. Luzkovich, Joseph J., et al., Springer (1995): 143–55.

18 Howitt, Michael R., et al., 'Tuft Cells, Taste-Chemosensory Cells, Orchestrate Parasite Type 2 Immunity in the Gut', *Science*, 351.6279 (2016): 1,329–33.

19 Verbeurgt, C., et al., 'The Human Bitter Taste Receptor T2R38 is Broadly Tuned for Bacterial Compounds', *PLoS One*, 12.9 (2017): e0181302.

20 Xu, J., et al., 'Functional Characterization of Bitter-Taste Receptors Expressed in Mammalian Testis', *MHR: Basic Science of Reproductive Medicine*, 19.1 (2012): 17–28.

21 Maurer, S., et al., 'Tasting Pseudomonas Aeruginosa Biofilms: Human Neutrophils Express the Bitter Receptor T2R38 as Sensor for the Quorum Sensing Molecule N-(3-oxododecanoyl)-l-homoserine lactone', *Frontiers in Immunology*, 6 (2015): 369, https://doi.org/10.3389/fimmu.2015.00369

22 Lin, W., et al., 'Epithelial Na+ Channel Subunits in Rat Taste Cells: Localization and Regulation by Aldosterone', *Journal of Comparative Neurology*, 405.3 (1999): 406–20; Pimenta, E., Gordon, R. D., and Stowasser, M., 'Salt, Aldosterone and Hypertension', *Journal of Human Hypertension*, 27.1 (2013): 1–6.

23 Rose, E. A., Porcerelli, J. H., and Neale, A. V., 'Pica: Common but Commonly Missed', *Journal of the American Board of Family Practice*, 13.5 (2000): 353–8.

24 Knaapila, Antti, et al., 'Genetic Analysis of Chemosensory Traits in Human Twins', *Chemical Senses*, 37.9 (2012): 869–81.

25 Dowd M., '"I'm President," So no more broccoli!', *New York Times*, 23 March 1990, http://www.nytimes.com/1990/03/23/us/i-m-president-so-no-more-broccoli.html; see also Hall, T., 'Broccoli, Hated by a President, is Capturing Popular Votes', *New York Times*, 25 March 1992, http://www.nytimes.com/1992/03/25/garden/broccoli-hated-by-a-president-is-capturing-popular-votes.html?pagewanted=all

26 Sandell, Mari A., and Breslin, Paul A. S., 'Variability in a Taste-Receptor Gene Determines Whether We Taste Toxins in Food', *Current Biology*, 16 (2006): R792–4.

27 Lipchock, S. V., et al., 'Human Bitter Perception Correlates With Bitter Receptor Messenger RNA Expression in Taste Cells', *American Journal of Clinical Nutrition*, 98. 4 (2013): 1,136–43.

28 See Bartoshuk, L. M., 'Comparing Sensory Experiences Across

Individuals: Recent Psychophysical Advances Illuminate Genetic Variation in Taste Perception', *Chemical Senses*, 25.4 (2000): 447–60.

29 Miller Jr, I. J., and Reedy Jr, F. E., 'Variations in Human Taste Bud Density and Taste Intensity Perception', *Physiology & Behavior*, 47.6 (1990): 1,213–19; for the test itself, see https://www.scientificamerican.com/article/super-tasting-science-find-out-if-youre-a-supertaster/

30 Masi, Camilla, et al., 'The Impact of Individual Variations in Taste Sensitivity on Coffee Perceptions and Preferences', *Physiology & Behavior*, 138 (2015): 219–26.

31 Lu, Ping, et al., 'Extraoral Bitter Taste Receptors in Health and Disease', *Journal of General Physiology*, 149.2 (2017): 181–97.

32 Adappa, Nithin D., et al., 'The Bitter Taste Receptor T2R38 is an Independent Risk Factor for Chronic Rhinosinusitis Requiring Sinus Surgery', *International Forum of Allergy & Rhinology*, 4.1 (2014): 3–7.

33 Choi, Jeong-Hwa, et al., 'Genetic Variation in the TAS2R38 Bitter Taste Receptor and Gastric Cancer Risk in Koreans', *Scientific Reports*, 6.1 (2016): 1–8.

34 Reed, Danielle R., and McDaniel, Amanda H., 'The Human Sweet Tooth', *BMC Oral Health*, 6.1 (2006), https://doi.org/10.1186/1472-6831-6-S1-S17

35 Mainland, Joel D., and Matsunami, Hiroaki, 'Taste Perception: How Sweet It Is (To Be Transcribed by You)', *Current Biology*, 19.15 (2009): R655–6.

36 Haznedaroğlu, Eda, et al., 'Association of Sweet Taste Receptor Gene Polymorphisms With Dental Caries Experience in School Children', *Caries Research*, 49.3 (2015): 275–81.

37 Raliou, M., Wiencis, A., et al., 'Nonsynonymous Single Nucleotide Polymorphisms in Human tas1r1, tas1r3, and mGluR1 and Individual Taste Sensitivity to Glutamate', *American Journal of Clinical Nutrition*, 90.3 (2009): 789S–799S.

38 Sagioglou, Christina, and Greitemeyer, Tobias, 'Individual Differences in Bitter Taste Preferences Are Associated With Antisocial Personality Traits', *Appetite*, 96 (2016): 299–308.

39 Sagioglou, Christina, and Greitemeyer, Tobias, 'Bitter Taste Causes Hostility', *Personality and Social Psychology Bulletin*, 40.12 (2014): 1,589–97.

40 Eskine, Kendall J., Kacinik, Natalie A., and Prinz, Jesse J., 'A Bad Taste in the Mouth: Gustatory Disgust Influences Moral Judgment', *Psychological Science*, 22.3 (2011): 295–9.

41 Ruskin, J., *Traffic*, Penguin Classics (2015).

42 Chapman, Hanah A., et al., 'In Bad Taste: Evidence for the Oral Origins of Moral Disgust', *Science*, 323.5918 (2009): 1,222–6.

43 Ren, Dongning, et al., 'Sweet Love: The Effects of Sweet Taste Experience on Romantic Perceptions', *Journal of Social and Personal Relationships*, 32.7 (2015): 905–21.

44 Wang, Liusheng, et al., 'The Effect of Sweet Taste on Romantic Semantic Processing: An ERP Study', *Frontiers in Psychology*, 10 (2019), https://doi.org/10.3389/fpsyg.2019.01573

45 Spence, C. *Gastrophysics: The New Science of Eating*, Viking (2017).

46 Velasco, Carlos, et al., 'Colour–Taste Correspondences: Designing Food Experiences to Meet Expectations or to Surprise', *International Journal of Food Design*, 1.2 (2016): 83–102.

47 See, for example, Spence, C., and Parise, C. V., 'The Cognitive Neuroscience of Crossmodal Correspondences', *i-Perception*, 3.7. (2012): 410–12.

48 Sievers, Beau, et al., 'A Multi-Sensory Code for Emotional Arousal', *Proceedings of the Royal Society B*, 286.1906 (2019), https://doi.org/10.1098/rspb.2019.0513; https://digest.bps.org.uk/2019/08/28/heres-why-spiky-shapes-seem-angry-and-round-sounds-are-calming/

49 Morrot, Gil, Brochet, Frédéric, and Dubourdieu, Denis, 'The Color of Odors', *Brain and Language*, 79.2 (2001): 309–20; Spence, C., 'The Colour of Wine – Part 1', *World of Fine Wine*, 28 (2010): 122–9.

50 Kaufman, Andrew, et al., 'Inflammation Arising From Obesity Reduces Taste Bud Abundance and Inhibits Renewal', *PLoS Biology*, 16.3 (2018): e2001959; Majid, A., and Levinson, S. C., 'Language Does Provide Support for Basic Tastes', *Behavioral and Brain Sciences*, 31.1 (2008): 86–7.

Chapter 5: Touch

1 Böhm, Jennifer, et al., 'The Venus Flytrap *Dionaea muscipula* Counts Prey-Induced Action Potentials to Induce Sodium Uptake', *Current Biology*, 26.3 (2016): R286–95.

2 Müller, J., trans. Baly, W. M., *Elements of Physiology*, vol. 2, Lea and Blanchard (1843).

3 Purves, D., et al., 'Mechanoreceptors Specialized to Receive Tactile Information', *Neuroscience* (2001).

4 Bell, Jonathan, Bolanowski, Stanley, and Holmes, Mark H., 'The Structure and Function of Pacinian Corpuscles: A Review', *Progress in Neurobiology*, 42.1 (1994): 79–128.

5 Miller, L. E., et al., 'Sensing With Tools Extends Somatosensory Processing Beyond the Body', *Nature*, 561.7722 (2018): 239–42.

6 Abraira, Victoria E., and Ginty, David D., 'The Sensory Neurons of Touch', *Neuron*, 79.4 (2013): 618–39.

7 https://faculty.washington.edu/chudler/receptor.html

8 Maksimovic, Srdjan, et al., 'Epidermal Merkel Cells are Mechanosensory Cells that Tune Mammalian Touch Receptors', *Nature*, 509.7502 (2014): 617–21.

9 Merkel, F., 'Tastzellen und Tastkörperchen bei den Hausthieren und beim Menschen', *Archiv für mikroskopische Anatomie*, 11. 1 (1875): 636–52.

10 Hoffman, B. U., et al., 'Merkel Cells Activate Sensory Neural Pathways Through Adrenergic Synapses', *Neuron*, 100. 6 (2018): 1,401–13.

11 Linden, David, J., *Touch: The Science of the Sense that Makes us Human*, Viking (2015).

12 Carpenter, Cody W., et al., 'Human Ability to Discriminate Surface Chemistry by Touch', *Materials Horizons*, 5.1 (2018): 70–7.

13 Lieber, J. D., and Bensmaia, S. J., 'High-Dimensional Representation of Texture in Somatosensory Cortex of Primates', *Proceedings of the National Academy of Sciences*, 116.8 (2019): 3,268–77.

14 https://www.illusionsindex.org/i/aristotle

15 Cicmil, N., Meyer, A. P., and Stein, J. F., 'Tactile Toe Agnosia and Percept of a "Missing Toe" in Healthy Humans', *Perception*, 45.3 (2016): 265–80.

16 http://www.ox.ac.uk/news/2015-09-22-confusion-afoot

17 See eg: Ackerley, R., et al., 'Touch Perceptions Across Skin Sites: Differences Between Sensitivity, Direction Discrimination and Pleasantness', *Frontiers in Behavioral Neuroscience*, 8. 54 (2014), https://doi.org/10.3389/fnbeh.2014.00054

18 Ackerley, Rochelle, et al., 'Human C-Tactile Afferents Are Tuned to the Temperature of a Skin-Stroking Caress', *Journal of Neuroscience*, 34.8 (2014): 2,879–83.

19 Vallbo, Å. B., Olausson, Hakan, and Wessberg, Johan, 'Unmyelinated Afferents Constitute a Second System Coding Tactile Stimuli of the Human Hairy Skin', *Journal of Neurophysiology*, 81.6 (1999): 2,753–63.

20 McGlone, Francis, Wessberg, Johan, and Olausson, Håkan, 'Discriminative and Affective Touch: Sensing and Feeling', *Neuron*, 82.4 (2014): 737–55.

21 https://gupea.ub.gu.se/handle/2077/51879

22 Field, Tiffany M., et al. 'Tactile/Kinesthetic Stimulation Effects on Preterm Neonates', *Pediatrics*, 77.5 (1986): 654–8.

23 Frenzel, Henning, et al., 'A Genetic Basis for Mechanosensory Traits in Humans', *PLoS Biology*, 10.5 (2012), https://doi.org/10.1371/journal.pbio.1001318

24 Ranade, S. S., et al., 'Piezo2 is the Major Transducer of Mechanical Forces for Touch Sensation in Mice', *Nature*, 516.7529 (2014): 121–5.

25 Chesler, A. T., et al., 'The Role of PIEZO2 in Human Mechanosensation', *New England Journal of Medicine*, 375.14 (2016): 1,355–64.

26 Harrar, Vanessa, Spence, Charles, and Makin, Tamar R., 'Topographic Generalization of Tactile Perceptual Learning', *Journal of Experimental Psychology: Human Perception and Performance*, 40.1 (2014): 15–23.

27 Muret, D. et al., 'Neuromagnetic Correlates of Adaptive Plasticity Across the Hand-Face Border in Human Primary Somatosensory Cortex.', J. *Neurophysiol.*, 115 (2016): 2,095–104.

28 Field, Tiffany, *Touch*, MIT Press, (2014).

29 From questionnaire responses gathered in 2018 and emailed to the author. BitterSuite's work has not been published.

30 Field, T. 'American Adolescents Touch Each Other Less and Are More Aggressive Toward Their Peers as Compared With French Adolescents', *Adolescence*, 34.136 (1999): 753–8.

31 https://greatergood.berkeley.edu/article/item/why_physical_touch_matters_for_your_well_being

32 Sonar, Harshal Arun, and Paik, Jamie, 'Soft Pneumatic Actuator Skin with Piezoelectric Sensors for Vibrotactile Feedback', *Frontiers in Robotics and AI*, 2 (2016), https://doi.org/10.3389/frobt.2015.00038

33 Tee, B. C.-K., et al., 'A skin-Inspired Organic Digital Mechanoreceptor', *Science*, 350.6258 (2015): 313–16.

34 Kim, Y., Chortos, et al., 'A Bioinspired Flexible Organic Artificial Afferent Nerve', *Science*, 360.6392 (2018): 998–1,003.

35 Ptito, Maurice, et al., 'Cross-Modal Plasticity Revealed by Electrotactile Stimulation of the Tongue in the Congenitally Blind', *Brain*, 128.3 (2005), 606–14; for a wonderful story on this, see Twilley, N., 'Seeing With Your Tongue', *New Yorker*, 8 May 2017, https://www.newyorker.com/magazine/2017/05/15/seeing-with-your-tongue

36 https://www.smithsonianmag.com/innovation/could-this-futuristic-vest-give-us-sixth-sense-180968852/; see also Neosensory: https://neosensory.com

Part Two

1 Sloan, Phillip Reid, ed., *The Hunterian Lectures in Comparative Anatomy (May and June 1837)*, University of Chicago Press (1992).

Chapter 6: Body Mapping

1 Pearce, J. M. S., 'Henry Charlton Bastian (1837–1915): Neglected Neurologist and Scientist', *European Neurology*, 63.2 (2010): 73–8.

2 See Liddell, Edward George Tandy, 'Charles Scott Sherrington 1857–1952', *Obituary Notices of Fellows of the Royal Society*, 8.21 (1952): 241–70.

3 Sherrington, Charles, 'The Integrative Action of the Nervous System', *Journal of Nervous and Mental Disease*, 34.12 (1907): 801–2; and see Burke, Robert E., 'Sir Charles Sherrington's the Integrative Action of the Nervous System: A Centenary Appreciation', *Brain*, 130.4 (2007): 887–94.

4 Sherrington, Charles, 'The Integrative Action of the Nervous System', *Journal of Nervous and Mental Disease*, 34.12 (1907): 801–2.

5 Sarmadi, Alireza, Sharbafi, Maziar Ahamd, and Seyfarth, André, 'Reflex Control of Body Posture in Standing', *2017 IEEE-RAS 17th International Conference on Humanoid Robotics (Humanoids)*, IEEE, 2017.

6 Sherrington, C. S., *Yale University Mrs. Hepsa Ely Silliman Memorial Lectures. The Integrative Action of the Nervous System* (1906), https://doi.org/10.1037/13798-000

7 Purves, Dale et al., eds, *Neuroscience*, 2nd edn, Sinauer Associates (2001).

8 Eccles, John Carew, 'Letters from CS Sherrington, FRS, to Angelo Ruffini between 1896 and 1903', *Notes and Records of the Royal Society of London*, 30.1 (1975): 69–88.

9 Eccles, J. C., 'Letters from C. S. Sherrington, F. R. S., to Angelo Ruffini Between 1896 and 1903', *Notes and Records of the Royal Society of London*, 30.1 (1975): 69–88

10 Gilman, S. 'Joint Position Sense and Vibration Sense: Anatomical Organisation and Assessment', *Journal of Neurology, Neurosurgery & Psychiatry*, 73.5 (2002): 473–7.

11 Gandevia, Simon C., et al., 'Motor Commands Contribute to Human Position Sense', *Journal of Physiology*, 571.3 (2006): 703–10.

12 Oby, E. R., Golub, et al., 'New Neural Activity Patterns Emerge With Long-Term Learning', *Proceedings of the National Academy of Sciences*, 116.30 (2019): 15,210–15.

13 https://thebrain.mcgill.ca/flash/i/i_03/i_03_cl/i_03_cl_dou/i_03_cl_dou.html

14 Cole, Jonathan, *Pride and a Daily Marathon*, MIT Press (1995); see also McNeill, David, Quaeghebeur, Liesbet, and Duncan, Susan, 'IW – "The Man Who Lost His Body"', *Handbook of Phenomenology and Cognitive Science*, Springer (2010): 519–43.

15 Woo, Seung-Hyun, et al., 'Piezo2 is the Principal Mechanotransduction Channel for Proprioception', *Nature Neuroscience*, 18.12 (2015): 1,756–62.

16 Mehring, C., et al., 'Augmented Manipulation Ability in Humans With Six-Fingered Hands', *Nature Communications*, 10.2401 (2019), https://doi.org/10.1038/s41467-019-10306-w

17 https://www.youtube.com/watch?v=Ks-_Mh1QhMc

18 See https://digest.bps.org.uk/2018/03/28/54-study-analysis-says-power-posing-does-affect-peoples-emotions-and-is-worth-researching-further/

19 https://www.telegraph.co.uk/rugby-union/2017/08/24/leading-haka-fires-like-adrenalin-rush/

20 https://www.sciencedaily.com/releases/2017/08/170801144247.htm; see also Liu, Y., and Medina, J., 'Influence of the Body Schema on Multisensory Integration: Evidence From the Mirror Box Illusion', *Scientific Reports*, 7.1 (2017):1–11.

21 Botvinick, Matthew, and Cohen, Jonathan, 'Rubber Hands "Feel" Touch that Eyes See', *Nature*, 391 (1998): 756.

22 https://www.tinyurl.com/hebarbie

23 Van Der Hoort, B., Guterstam, A., and Ehrsson, H. H., 'Being Barbie: The Size of One's Own Body Determines the Perceived Size of the World', *PloS ONE*, 6.5 (2011): e20195.

24 Michel, Charles, et al., 'The Butcher's Tongue Illusion', *Perception*, 43.8 (2014): 818–24.

25 Sutton, J., 'Interview: "People have been ignoring the body for a long time"', *Psychologist*, 27 (March 2014): 177–8, https://thepsychologist.bps.org.uk/volume-27/edition-3/interview-people-have-been-ignoring-body-long-time

26 Moseley, G. Lorimer, et al., 'Psychologically Induced Cooling of a

Specific Body Part Caused by the Illusory Ownership of an Artificial Counterpart', *Proceedings of the National Academy of Sciences*, 105.35 (2008): 13,169–73.

27 Barnsley, N., et al., 'The Rubber Hand Illusion Increases Histamine Reactivity in the Real Arm', *Current Biology*, 21.23 (2011): R945–6.

28 Dieter, Kevin C., et al., 'Kinesthesis Can Make an Invisible Hand Visible', *Psychological Science*, 25.1 (2014): 66–75.

29 Fagard, J., et al., 'Fetal Origin of Sensorimotor Behavior', *Frontiers in Neurorobotics*, 12.23 (2018), https://doi.org/10.3389/fnbot.2018.00023

30 Howes, David, and Classen, Constance, 'Sounding Sensory Profiles', *Epilogue to The Varieties of Sensory Experience Howes*, David, ed, University of Toronto Press (1991).

31 Shubert, Tiffany E., et al. 'The Effect of an Exercise-Based Balance Intervention on physical and Cognitive Performance for Older Adults: A Pilot Study', *Journal of Geriatric Physical Therapy* 33.4 (2010): 157–64; Alloway, Ross G., and Alloway, Tracy Packiam, 'The Working Memory Benefits of Proprioceptively Demanding Training: A Pilot Study', *Perceptual and Motor Skills*, 120.3 (2015): 766–75.

32 Ribeiro, Fernando, and Oliveira, José, 'Aging Effects on Joint Proprioception: The Role of Physical Activity in Proprioception Preservation', *European Review of Aging and Physical Activity*, 4.2 (2007): 71–76.

33 Liu, Jing, et al., 'Effects of Tai Chi Versus Proprioception Exercise Program on Neuromuscular Function of the Ankle in Elderly People: A Randomized Controlled Trial', *Evidence-based Complementary and Alternative Medicine* (2012), https://doi.org/10.1155/2012/265486

34 Fritzsch, Bernd, Kopecky, Benjamin J., and Duncan, Jeremy S., 'Development of the Mammalian "vestibular" System: Evolution of Form to Detect Angular and Gravity Acceleration', *Development of Auditory and Vestibular Systems*, Academic Press (2014): 339–67.

Chapter 7: Gravity and Whole-body Motion

1 Day, Brian, and Fitzpatrick, Richard C., 'The Vestibular System', *Current Biology*, 15.15 (2005): R583–6.

2 Loftus, Brian D., et al., in *Neurology Secrets*, 5th edn, ed. Rolak, Loren A., Mosby/Elsevier (2010).

3 Romand, Raymond, and Varela-Nieto, Isabel, eds, *Development of Auditory and Vestibular Systems*, Academic Press (2014), Chapter 12.

4 Solé, M., et al., 'Does Exposure to Noise From Human Activities Compromise Sensory Information From Cephalopod Statocysts?', *Deep Sea Research Part II: Topical Studies in Oceanography*, 95 (2013): 160–81.

5 Franklin, C. L., 'An Unknown Organ of Sense', *Science*, 14.345 (1889): 183–5.

6 https://www.newyorker.com/magazine/1999/04/05/the-man-who-walks-on-air

7 https://www.theguardian.com/sport/video/2014/nov/03/nik-wallenda-skyscraper-tightrope-blindfold-twice-video

8 Hippocrates, trans. Jones, W. H. S., *Hippocrates Volume IV*, Loeb Classical Library 150 (1931). (The work in this collection is not necessarily attributed to Hippocrates himself, but to his tradition.)

9 Kennedy, Robert S., et al., 'Symptomatology Under Storm Conditions in the North Atlantic in Control Subjects and in Persons with Bilateral Labyrinthine Defects', *Acta oto-laryngologica*, 66.1–6 (1968): 533–40.

10 Scherer, H., et al., 'On the Origin of Interindividual Susceptibility to Motion Sickness', *Acta oto-laryngologica*, 117.2 (1997): 149–53.

11 Perrault, Aurore A., et al., 'Whole-Night Continuous Rocking Entrains Spontaneous Neural Oscillations With Benefits for Sleep and Memory', *Current Biology*, 29.3 (2019): R402–11.

12 Pasquier, Florane, et al., 'Impact of Galvanic Vestibular Stimulation on Anxiety Level in Young Adults', *Frontiers in Systems Neuroscience*, 13 (2019), https://doi.org/10.3389/fnsys.2019.00014

13 https://ich.unesco.org/en/RL/mevlevi-sema-ceremony-00100; http://mevlanafoundation.com/mevlevi_order_en.html

14 Cakmak, Y. O., et al., 'A Possible Role of Prolonged Whirling Episodes on Structural Plasticity of the Cortical Networks and Altered Vertigo Perception: The Cortex of Sufi Whirling Dervishes', *Frontiers in Human Neuroscience* (2017), https://doi.org/10.3389/fnhum.2017.00003

15 Lopez, Christophe, and Elzière, Maya, 'Out-of-Body Experience in Vestibular Disorders – A Prospective Study of 210 Patients With Dizziness', *Cortex*, 104 (2018): 193–206.

16 Blanke, Olaf, et al., 'Stimulating Illusory Own-Body Perceptions', *Nature* 419.6904 (2002): 269–70.

17 Tianwu, H., et al., 'Effects of Alcohol Ingestion on Vestibular Function in Postural Control', *Acta Oto-Laryngologica*, 115.519 (1995): 127–31; see also, Shibano, Stacie, 'The Vestibular System and the "Spins": A Proposal' (2013), http://greymattersjournal.com/the-vestibular-system-and-the-spins-a-proposal

18 Rosenberg, Marissa J., et al., 'Human Manual Control Precision Depends on Vestibular Sensory Precision and Gravitational Magnitude', *Journal of Neurophysiology*, 120.6 (2018): 3,187–97.

19 Bermúdez Rey, M. C., et al., 'Vestibular Perceptual Thresholds Increase Above the Age of 40', *Frontiers in Neurology* (2016), https://doi.org/10.3389/fneur.2016.00162

20 Ibid.

21 https://www.sciencedaily.com/releases/2016/11/161128085345.htm

22 Ibid.

23 Agrawal, Y., et al., 'Disorders of Balance and Vestibular Function in US Adults: Data from the National Health and Nutrition Examination Survey, 2001–2004', *Archives of Internal Medicine*, 169.10 (2009): 938–44.

24 Serrador, Jorge M., et al., 'Vestibular Effects on Cerebral Blood Flow', *BMC Neuroscience*, 10.119 (2009), https://doi:10.1186/1471-2202-10-119

Chapter 8: Inner-sensing

1 https://www.health.harvard.edu/staying-healthy/understanding-the-stress-response; http://mcb.berkeley.edu/courses/mcb160/Fall2005Slides/Wk12F_111805.pdf

2 See Holmes, F. L., 'Claude Bernard, The "Milieu Intérieur", and Regulatory Physiology', *History and Philosophy of the Life Sciences*, 8.1 (1986): 3–25.

3 Cannon, Walter, 'Organization for Physiological Homeostasis', *Physiological Reviews*, 9:3 (1929): 399–431; see also, Cooper, S. J., 'From Claude Bernard to Walter Cannon: Emergence of the Concept of Homeostasis', *Appetite*, 51.3 (2008): 419–27.

4 Sherrington, C., *The Integrative Action of the Nervous System*, Scribner (1906).

5 Nonomura, Keiko, et al., 'Piezo2 Senses Airway Stretch and Mediates Lung Inflation-Induced Apnoea', *Nature*, 541.7636 (2017): 176–81.

6 Parkes, M. J., 'Breath-Holding and Its Breakpoint', *Experimental Physiology*, 91.1 (2006): 1–15.

7 de Wolf, Elizabeth, Cook, Jonathan, and Dale, Nicholas, 'Evolutionary Adaptation of the Sensitivity of Connexin26 Hemichannels to CO2', *Proceedings of the Royal Society B: Biological Sciences*, 284 (1848) (2017), https://doi.org/10.1098/rspb.2016.2723; see also Jalalvand, Elham, et al., 'Cerebrospinal Fluid-Contacting Neurons Sense pH Changes and

Motion in the Hypothalamus', *Journal of Neuroscience* 38.35 (2018): 7,713–24.

8   Cannon, W. B., 'Physiological Regulation of Normal States: Some Tentative Postulates Concerning Biological Homeostatics', Editions Médicales (1926).

9   Yuan, Guoxiang, et al., 'Protein Kinase G–Regulated Production of H2S Governs Oxygen Sensing', *Sci. Signal*, 8.373 (2015): ra37–ra37.

10 Chapleau, M. W., 'Cardiovascular Mechanoreceptors', in Ito, F., ed., *Comparative Aspects of Mechanoreceptor Systems: Advances in Comparative and Environmental Physiology*, 10 (1992): 137–164.

11 Zeng, Wei-Zheng, et al., 'PIEZOs Mediate Neuronal Sensing of Blood Pressure and the Baroreceptor Reflex', Science 362.6413 (2018): 464–7; Xu, Jie, et al., 'GPR68 Senses Flow and is Essential for Vascular Physiology', *Cell*, 173.3 (2018): 762–75.

12 Most of the detail about Nitsch and his dives comes from personal conversations, but for more, see www.herbertnitsch.com and also the documentary *Herbert Nitsch, Back from the Abyss* (2013).

13 Garfinkel, S. N., et al., 'Knowing Your Own Heart: Distinguishing Interoceptive Accuracy From Interoceptive Awareness', *Biological Psychology*, 104 (2015): 65–74.

14 Herbert, Beate M., Ulbrich, Pamela, and Schandry, Rainer, 'Interoceptive Sensitivity and Physical Effort: Implications for the Self-Control of Physical Load in Everyday Life', *Psychophysiology*, 44.2 (2007): 194–202.

15 For more detail on this, see, for example, Koch, A., and Pollatos, O., 'Interoceptive Sensitivity, Body Weight and Eating Behavior in Children: A Prospective Study', *Frontiers in Psychology*, 5 (2014), https://doi.org/10.3389/fpsyg.2014.01003; Herbert, Beate M., and Pollatos, Olga. 'Attenuated Interoceptive Sensitivity in Overweight and Obese Individuals', *Eating Behaviors*, 15.3 (2014): 445–8.

16 Critchley, Hugo D., and Harrison, Neil A., 'Visceral Influences on Brain and Behavior', *Neuron*, 77.4 (2013): 624–38.

17 See, for example, Porges, S. W., 'Cardiac Vagal Tone: A Physiological Index of Stress', *Neuroscience & Biobehavioral Reviews*, 19.2 (1995): 225–33.

18 For more on vagal tone and health, see Young, Emma, 'Vagus Thinking: Meditate Your Way to Better Health', *New Scientist*, 10 July 2013.

19 Hansen, Anita Lill, et al., 'Heart Rate Variability and Its Relation to Prefrontal Cognitive Function: The Effects of Training and Detraining', *European Journal of Applied Physiology*, 93.3 (2004): 263–72.

20 Thayer, Julian F., and Lane Richard D., 'The Role of Vagal Function in the Risk for Cardiovascular Disease and Mortality', *Biological Psychology*, 74.2 (2007): 224–42.

21 See Vince, Gaia, 'Hacking the Nervous System', *Mosaic*, 25 May 2015.

22 Oveis, Christopher, et al., 'Resting Respiratory Sinus Arrhythmia is Associated with Tonic Positive Emotionality', *Emotion*, 9.2 (2009): 265–270.

23 Hansen, Anita Lill, Johnsen, Bjørn Helge, and Thayer, Julian F., 'Vagal Influence on Working Memory and Attention', *International Journal of Psychophysiology*, 48.3 (2003): 263–74.

24 Guiraud, Thibaut, et al., 'High-Intensity Interval Exercise Improves Vagal Tone and Decreases Arrhythmias in Chronic Heart Failure', *Medicine & Science in Sports & Exercise*, 45.10 (2013): 1,861–7.

Chapter 9: Temperature

1 Fairclough, Stephen, and King, Nicole, 'Choanoflagellates: Choanoflagellida, Collared-Flagellates' (14 August 2006), https://tolweb.org/Choanoflagellates/2375/2006.08.14 in the 'Tree of Life Web Project'

2 Wang, H., and Siemens, J., 'TRP Ion Channels in Thermosensation, Thermoregulation and Metabolism', *Temperature*, 2.2 (2015): 178–87.

3 Moparthi, L., et al., 'Human TRPA1 is Intrinsically Cold and Chemosensitive With and Without Its N-terminal Ankyrin Repeat Domain', *Proceedings of the National Academy of Sciences*, 111.47 (2014): 16,901–6; Myers, B. R., Sigal, Y. M., and Julius, D., 'Evolution of Thermal Response Properties in a Cold-Activated TRP Channel', *PloS ONE*, 4.5 (2009): e5741; Bautista, D. M., et al., 'The Menthol Receptor TRPM8 is the Principal Detector of Environmental Cold', *Nature*, 448.7150 (2007): 204–8.

4 Kraft, K. H., et al., 'Multiple Lines of Evidence for the Origin of Domesticated Chili Pepper, Capsicum annuum, in Mexico', *Proceedings of the National Academy of Sciences*, 111.17 (2014): 6,165–70.

5 Han, Y., Li, B., et al., 'Molecular Mechanism of the Tree Shrew's Insensitivity to Spiciness', *PLoS Biology*, 16.7 (2018): e2004921.

6 Siemens, J., et al., 'Spider Toxins Activate the Capsaicin Receptor to Produce Inflammatory Pain', *Nature*, 444.7116 (2006): 208–12.

7 http://blog.monell.org/02/22/introducing-marco-tizzano/

8 Smith, C. U. M., *Biology of Sensory Systems*, Wiley-Blackwell (2008) (also for the sections below).

9 Morrison, S. F., 'Central Control of Body Temperature', *F1000Research*, 5 (2016), https://doi.org/10.12688/f1000research.7958.1

10 Stevens, K. C., and Choo, K. K., 'Temperature Sensitivity of the Body Surface Over the Life Span', *Somatosensory & Motor Research*, 15.1 (1998): 13–28.

11 https://www.heart.co.uk/showbiz/celebrities/definitive-list-worlds-sexiest-men-2020/

12 Conference report: https://www.newscientist.com/article/dn10213-women-become-sexually-aroused-as-quickly-as-men/

13 Stevens, J. C., and Green, B. G., 'Temperature–Touch interaction: Weber's Phenomenon Revisited', *Sensory Processes*, 2.3 (1978): 206–219.

14 Frankmann, S. P., and Green, B. G., 'Differential Effects of Cooling on the Intensity of Taste', *NYASA*, 510.1 (1987): 300–3.

15 Green, B. G., Lederman, S. J., and Stevens, J. C., 'The Effect of Skin Temperature on the Perception of Roughness', *Sensory Processes*, 3.4 (1979): 327–33.

16 Stevens, J. C., 'Temperature Can Sharpen Tactile Acuity', *Perception & Psychophysics*, 31.6 (1982): 577–80.

17 Gröger, Udo, and Wiegrebe, Lutz, 'Classification of Human Breathing Sounds by the Common Vampire Bat, Desmodus rotundus', *BMC Biology*, 4.1 (2006): 18.

18 Gracheva, Elena O., et al., 'Ganglion-Specific Splicing of TRPV1 Underlies Infrared Sensation in Vampire Bats', *Nature*, 476.7358 (2011): 88–91.

19 Story, Gina M., 'The Emerging Role of TRP Channels in Mechanisms of Temperature and Pain Sensation', *Current Neuropharmacology*, 4.3 (2006): 183–96.

20 See https://www.ncbi.nlm.nih.gov/gene/7442; Xu, H., et al., 'Functional Effects of Nonsynonymous Polymorphisms in the Human TRPV1 Gene', *American Journal of Physiology-Renal Physiology*, 293.6 (2007): F1865–76.

21 https://www.guinnessworldrecords.com/world-records/hottest-chili

22 Spinney, J., 'Consciousness Isn't Just the Brain', *New Scientist* (24 June 2020).

23 https://www.ocregister.com/2016/09/30/how-to-survive-eating-a-carolina-reaper-the-worlds-hottest-pepper/

24 Gianfaldoni, Serena, et al., 'History of the Baths and Thermal

Medicine', *Open Access Macedonian Journal of Medical Sciences*, 5.4 (2017): 566–568.

25 Fagan, Garrett, C., *Bathing in Public in the Roman World*, University of Michigan Press (2002).

26 Zaccardi, F., et al., 'Sauna Bathing and Incident Hypertension: A Prospective Cohort Study', *American Journal of Hypertension*, 30.11 (2017): 1,120–5.

27 Cochrane, Darryl J., 'Alternating Hot and Cold Water Immersion for Athlete Recovery: A Review', *Physical Therapy in Sport*, 5.1 (2004): 26–32.

28 https://www.bbc.co.uk/sport/tennis/40489130

29 Chang, T. Y., and Kajackaite, A., 'Battle for the Thermostat: Gender and the Effect of Temperature on Cognitive Performance', *PloS ONE*, 14.5 (2019): e0216362.

30 Pliny the Elder, *Natural History*, https://doi.org/10.4159/DLCL.pliny_elder-natural_history.1938

31 Moussaieff, A., et al., 'Incensole Acetate, an Incense Component, Elicits Psychoactivity by Activating TRPV3 Channels in the Brain', *FASEB Journal*, 22.8 (2008): 3,024–34.

32 Bargh, J. A., and Shalev, I., 'The Substitutability of Physical and Social Warmth in Daily Life', *Emotion*, 12.1 (2012): 154–62.

33 https://digest.bps.org.uk/2020/01/27/cold-days-can-make-us-long-for-social-contact-but-warming-up-our-bodies-eliminates-this-desire/

Chapter 10: Pain

1 In 2019, the International Association for the Study of Pain (IASP) proposed a new definition of pain as 'an unpleasant sensory and emotional experience associated with actual or potential tissue damage, or described in terms of such damage'.

2 Descartes, *Treatise of Man*, Prometheus Books (2003).

3 Sherrington, C. S., 'Qualitative Differences of Spinal Reflex Corresponding with Qualitative Difference of Cutaneous Stimulus', *Journal of Physiology*, 30 (1903): 39–46.

4 '50 Shades of Pain', *Nature*, 535.200 (14 July 2016), https://doi.org/10.1038/535200a

5 Dubin, Adrienne E., and Patapoutian, Ardem, 'Nociceptors: The Sensors of the Pain Pathway', *Journal of Clinical Investigation*, 120.11 (2010): 3,760–72.

6 Tracey Jr, W. Daniel, 'Nociception', *Current Biology*, 27.4 (2017): R129–33.

7 Jones, Nicholas G., et al., 'Acid-Induced Pain and Its Modulation in Humans', *Journal of Neuroscience*, 24.48 (2004): 10,974–9.

8 Bryant, Bruce P., 'Mechanisms of Somatosensory Neuronal Sensitivity to Alkaline pH', *Chemical Senses*, 30.1 (2005): i196–7, https://doi.org/10.1093/chemse/bjh182

9 Rivlin, R. S., 'Historical Perspective on the Use of Garlic', *Journal of Nutrition*, 131.3 (2001): 951S–4S.

10 Sharp, O, Waseem, S., and Wong, K. Y., 'A Garlic Burn: BMJ Case Reports', *BMJ Case Reports* 2018, https://doi.org/10.1136/bcr-2018-226027

11 https://nba.uth.tmc.edu/neuroscience/m/s2/chapter06.html

12 Benly, P., 'Role of Histamine in Acute Inflammation', *Journal of Pharmaceutical Sciences and Research*, 7.6 (2015): 373–376.

13 Han, Liang, et al., 'A Subpopulation of Nociceptors Specifically Linked to Itch', *Nature Neuroscience* 16.2 (2013): 174–182.

14 Benly, P., 'Role of Histamine in Acute Inflammation', *Journal of Pharmaceutical Sciences and Research*, 7.6 (2015): 373–376.

15 https://www.ucl.ac.uk/anaesthesia/sites/anaesthesia/files/PainPathwaysIntroduction.pdf

16 Wager, T. D., et al., 'An fMRI-Based Neurologic Signature of Physical Pain', *New England Journal of Medicine*, 368.15 (2013): 1,388–97.

17 https://mrc.ukri.org/news/blog/painless-a-q-a-with-geoff-woods/?redirected-from-wordpress

18 Ossipov, Michael H., Dussor, Gregory O., and Porreca, Frank, 'Central Modulation of Pain', *Journal of Clinical Investigation*, 120.11 (2010): 3,779–87.

19 Livingstone, David, *Missionary Travels and Researches in South Africa*, Chapter 1, https://www.gutenberg.org/files/1039/1039-h/1039-h.htm

20 https://www.theguardian.com/science/2019/mar/28/scientists-find-genetic-mutation-that-makes-woman-feel-no-pain

21 Critchley, H. D., and Garfinkel, S. N., 'Interactions Between Visceral Afferent Signaling and Stimulus Processing', *Frontiers in Neuroscience*, 9 (2015), https://doi.org/10.3389/fnins.2015.00286

22 https://www.shu.ac.uk/research/in-action/projects/vr-and-burns

23 Keltner, John R., et al., 'Isolating the Modulatory Effect of Expectation on Pain Transmission: A Functional Magnetic Resonance Imaging Study', *Journal of Neuroscience* 26.16 (2006): 4,437–43; Colloca, Luana, and

Benedetti, Fabrizio, 'Nocebo Hyperalgesia: How Anxiety is Turned Into Pain', *Current Opinion in Anesthesiology*, 20.5 (2007): 435–9.

24 Petrovic, Predrag, et al., 'Placebo and Opioid Analgesia – Imaging a Shared Neuronal Network', *Science*, 295.5560 (2002): 1,737–40.

25 Stephens, Richard, Atkins, John, and Kingston, Andrew, 'Swearing as a Response to Pain', *Neuroreport*, 20.12 (2009): 1,056–60.

26 https://thebrain.mcgill.ca/flash/i/i_03/i_03_cl/i_03_cl_dou/i_03_cl_dou.html

27 Goldstein, Pavel, Weissman-Fogel, Irit, and Shamay-Tsoory, Simone G., 'The Role of Touch in Regulating Inter-Partner Physiological Coupling During Empathy for Pain', *Scientific Reports*, 7.3252 (2017), https://doi.org/10.1038/s41598-017-03627-7

28 Lawler, Andrew, 'Did Ancient Mesopotamians Get High? Near Eastern Rituals May Have Included Opium, Cannabis', *Science* (2018), https://doi.org/ 10.1126/science.aat9271; Ren, Meng, et al., 'The Origins of Cannabis Smoking: Chemical Residue Evidence from the First Millennium BCE in the Pamirs', *Science Advances*, 5.6 (2019), https://doi.org/10.1126/sciadv.aaw1391

29 https://www.exeter.ac.uk/news/research/title_645441_en.html

30 DeWall, C. Nathan, et al., 'Acetaminophen Reduces Social Pain: Behavioral and Neural Evidence', *Psychological Science*, 21.7 (2010): 931–7.

31 Sznycer, Daniel, et al., 'Cross-Cultural Invariances in the Architecture of Shame', *Proceedings of the National Academy of Sciences*, 115.39 (2018): 9,702–7.

32 https://news.feinberg.northwestern.edu/2015/02/garcia-auditory-pathway/; Okamoto, Keiichiro, et al., 'Bright Light Activates a Trigeminal Nociceptive Pathway', *Pain*, 149.2 (2010): 235–42.

Chapter 11: Gut Feelings

1 Pankhurst, E., *My Own Story*, Eveleigh Nash (1914), https://www.gutenberg.org/files/34856/34856-h/34856-h.htm

2 https://www.parliament.uk/about/living-heritage/transformingsociety/electionsvoting/womenvote/overview/deedsnotwords/ ; https://www.theguardian.com/commentisfree/libertycentral/2009/jul/06/suffragette-hunger-strike-protest

3 https://spartacus-educational.com/Whunger.htm

4 Chantranupong, Lynne, Wolfson, Rachel L., and Sabatini, David M.,

'Nutrient-Sensing Mechanisms Across Evolution', *Cell*, 161.1 (2015): 67–83.

5 Osorio, Marina Borges, et al., 'SPX4 Acts on PHR1-Dependent and Independent Regulation of Shoot Phosphorus Status in Arabidopsis', *Plant Physiology*, 181.1 (2019): 332–52; Chien, Pei-Shan, et al., 'Sensing and Signaling of Phosphate Starvation: From Local to Long Distance', *Plant and Cell Physiology*, 59.9 (2018): 1,714–22.

6 Cannon, W. B., and Washburn, A. L., 'An Explanation of Hunger', *American Journal of Physiology*, 29 (1912): 441–54.

7 https://ourworldindata.org/hunger-and-undernourishment

8 Santacà, Maria, et al., 'Can Reptiles Perceive Visual Illusions? Delboeuf Illusion in Red-Footed Tortoise (Chelonoidis carbonaria) and Bearded Dragon (Pogona vitticeps)', *Journal of Comparative Psychology*, 133.4 (2019): 419–27.

9 Bai, L., et al., 'Genetic Identification of Vagal Sensory Neurons That Control Feeding', *Cell*, 179.5 (2019): 1,129–43.

10 Van Dyck, Zoé, et al., 'The Water Load Test as a Measure of Gastric Interoception: Development of a Two-Stage Protocol and Application to a Healthy Female Population', *PloS ONE*, 11.9 (2016), https://doi.org/10.1371/journal.pone.0163574

11 Cummings, D. E., and Overduin, J., 'Gastrointestinal Regulation of Food Intake', *Journal of Clinical Investigation*, 117.1 (2007): 13–23.

12 Herbert, B. M., et al., 'Interoception Across Modalities: On the Relationship Between Cardiac Awareness and the Sensitivity for Gastric Functions', *PloS ONE*, 7.5 (2012): e36646.

13 Koch, Anne, and Pollatos, Olga, 'Interoceptive Sensitivity, Body Weight and Eating Behavior in Children: A Prospective Study', *Frontiers in Psychology*, 5 (2014): 1,003.

14 https://digest.bps.org.uk/2017/12/13/imagining-bodily-states-like-feeling-full-can-affect-our-future-preferences-and-behaviour/

15 MacCormack, Jennifer K., and Lindquist, Kristen A., 'Feeling Hangry? When Hunger is Conceptualized as Emotion', *Emotion*, 19.2 (2019): 301–19.

16 Kalra, Priya B., Gabrieli, John D. E., and Finn, Amy S., 'Evidence of Stable Individual Differences in Implicit Learning', *Cognition* 190 (2019): 199–211.

17 Werner, Natalie S., et al., 'Enhanced Cardiac Perception is Associated with Benefits in Decision-Making', *Psychophysiology*, 46.6 (2009): 1,123–9.

18 Dunn, Barnaby D., et al., 'Listening to Your Heart: How Interoception Shapes Emotion Experience and Intuitive Decision Making', *Psychological Science*, 21.12 (2010): 1,835–44.

19 Kandasamy, Narayanan, et al., 'Interoceptive Ability Predicts Survival on a London Trading Floor', *Scientific Reports*, 6.1 (2016): 1–7.

20 Mitchell, H. H., et al., 'The Chemical Composition of the Adult Human Body and Its Bearing on the Biochemistry of Growth', *Journal of Biological Chemistry*, 158.3 (1945): 625–37.

21 Chumlea, W. Cameron, et al., 'Total Body Water Data for White Adults 18 to 64 Years of Age: The Fels Longitudinal Study', *Kidney International*, 56.1 (1999): 244–52.

22 Verbalis, Joseph G., 'How Does the Brain Sense Osmolality?', *Journal of the American Society of Nephrology*, 18.12 (2007): 3,056–9.

23 Zimmerman, Christopher A., et al., 'A Gut-to-Brain Signal of Fluid Osmolarity Controls Thirst Satiation', *Nature*, 568.7750 (2019): 98–102.

24 https://www.sciencedaily.com/releases/2019/03/190327142026.htm

25 Valtin, Heinz, and (With the Technical Assistance of Sheila A. Gorman),'"Drink at Least Eight Glasses of Water a Day." Really? Is There Scientific Evidence for "8×8"?', *American Journal of Physiology-Regulatory, Integrative and Comparative Physiology*, 283.5 (2002): R993–1004.

26 Saker, P., et al, 'Overdrinking, Swallowing Inhibition, and Regional Brain Responses Prior to Swallowing', *Proceedings of the National Academy of Sciences*, 113.43 (2016): 12,274–9.

27 https://www.sciencedaily.com/releases/2016/10/161007111027.htm

28 Miyamoto, Tatsuya, et al., 'Functional Role for Piezo1 in Stretch-Evoked Ca2+ Influx and ATP Release in Urothelial Cell Cultures', *Journal of Biological Chemistry*, 289.23 (2014): 16,565–75.

Part Three

1 The studies that I cite in this field do not tend to gather data on gender as distinct from biological sex. When a study reports on 'men' and 'women', the assumption is that the participants are the gender they were assigned at birth.

Chapter 12: A Sense of Direction

1 Ishikawa, Toru, and Montello, Daniel R., 'Spatial Knowledge Acquisition from Direct Experience in the Environment: Individual Differences in the Development of Metric Knowledge and the Integration of Separately Learned Places', *Cognitive Psychology*, 52.2 (2006): 93–129.

2 Hegarty, M., et al., 'Development of a Self-Report Measure of Environmental Spatial Ability', *Intelligence*, 30.5 (2002): 425–47.

3 O'Keefe, John, and Dostrovsky, Jonathan, 'The Hippocampus as a Spatial Map: Preliminary Evidence from Unit Activity in the Freely-Moving Rat', *Brain Research*, 34.1 (1971): 171–5.

4 Hafting, Torkel, et al., 'Microstructure of a Spatial Map in the Entorhinal Cortex', *Nature*, 436.7052 (2005): 801–6.

5 Epstein, Russell A., et al., 'The Cognitive Map in Humans: Spatial Navigation and Beyond', *Nature Neuroscience*, 20.11 (2017): 1,504–13.

6 Preston-Ferrer, Patricia, et al., 'Anatomical Organization of Presubicular Head-Direction Circuits', *eLife*, 5 (2016): e14592.

7 Vélez-Fort, Mateo, et al., 'A Circuit for Integration of Head-and Visual-Motion Signals in Layer 6 of Mouse Primary Visual Cortex', *Neuron*, 98.1 (2018): 179–91.

8 Guerra, Patrick A., Gegear, Robert J., and Reppert, Steven M., 'A Magnetic Compass Aids Monarch Butterfly Migration', *Nature Communications*, 5.1 (2014): 1–8; Gegear, Robert, J., et al., 'Demystifying Monarch Butterfly Migration', *Current Biology*, 28.17 (2018): R1009–22, https://doi.org/10.1016/j.cub.2018.02.067

9 Eder, Stephan H. K., et al., 'Magnetic Characterization of Isolated Candidate Vertebrate Magnetoreceptor Cells', *Proceedings of the National Academy of Sciences*, 109.30 (2012): 12,022–7.

10 Lohmann, Kenneth J., Putman, Nathan F., and Lohmann, Catherine M. F., 'Geomagnetic Imprinting: A Unifying Hypothesis of Long-Distance Natal Homing in Salmon and Sea Turtles', *Proceedings of the National Academy of Sciences*, 105.49 (2008): 19,096–101.

11 Gould, James L., 'Animal Navigation: The Evolution of Magnetic Orientation', *Current Biology*, 18.11 (2008): R482–4; Sutton, Gregory P., et al., 'Mechanosensory Hairs in Bumblebees (Bombus terrestris) Detect Weak Electric Fields', *Proceedings of the National Academy of Sciences*, 113.26 (2016): 7,261–5.

12 Chong, Lisa D., et al., 'Animal Magnetoreception', *Science*, 351.6278 (11 March 2016): 1,163–4.

13 Foley, Lauren E., Gegear, Robert J., and Reppert, Steven M., 'Human Cryptochrome Exhibits Light-Dependent Magnetosensitivity', *Nature Communications*, 2.1 (2011): 1–3.

14 Nießner, Christine, et al., 'Cryptochrome 1 in Retinal Cone Photoreceptors Suggests a Novel Functional Role in Mammals', *Scientific Reports*, 6 (2016), https://doi.org/10.1038/srep21848

15 Wang, Connie X., et al., 'Transduction of the Geomagnetic Field as Evidenced from Alpha-Band Activity in the Human Brain', *eNeuro*, 6.2 (2019), https://doi.org/10.1523/ENEURO.0483-18.2019

16 Jacobs, L. F., et al., 'Olfactory Orientation and Navigation in Humans', *PLoS ONE*, 10.6 (2015): e0129387.

17 Moser, May-Britt, Rowland, David C., and Moser, Edvard I., 'Place Cells, Grid Cells, and Memory', *Cold Spring Harbor Perspectives in Biology*, 7.2 (2015), https://doi.org/10.1101/cshperspect.a021808

18 Dahmani, Louisa, et al., 'An Intrinsic Association Between Olfactory Identification and Spatial Memory in Humans', *Nature Communications*, 9.1 (2018): 1–12.

19 https://jeb.biologists.org/content/222/Suppl_1/jeb186924

20 https://www.sciencedaily.com/releases/2015/06/150617175250.htm

21 Sharp, Andrew, 'Polynesian Navigation: Some Comments', *Journal of the Polynesian Society* (1963): 384–96.

22 Lewis, D., *The Voyaging Stars: Secrets of the Pacific Island Navigators* (1978).

23 Souman, J. L., et al., 'Walking Straight Into Circles', *Current Biology*, 19.18 (2009): R1,538–42.

24 Bestaven, Emma, Guillaud, Etienne, and Cazalets, Jean-René, 'Is "Circling" Behavior in Humans Related to Postural Asymmetry?', *PloS ONE*, 7.9 (2012), https://doi.org/10.1371/journal.pone.0043861

25 Young, Emma, 'The Disoriented Ape: Why Clever People Can be Terrible Navigators, *New Scientist* (12 December 2018).

26 Gagnon, K. T., et al., 'Sex Differences in Exploration Behavior and the Relationship to Harm Avoidance', *Human Nature*, 27.1 (2016): 82–97.

27 Cashdan, E., and Gaulin, S. J., 'Why Go There? Evolution of Mobility and Spatial Cognition in Women and Men', *Human Nature*, 27.1 (2016): 1–15.

28 Patai, E. Z., et al., 'Hippocampal and Retrosplenial Goal Distance Coding After Long-Term Consolidation of a Real-World Environment', *Cerebral Cortex*, 29.6 (2019): 2,748–58; Javadi, A. H., et al., 'Hippocampal and Prefrontal Processing of Network Topology to Simulate the Future', *Nature Communications*, 8.1 (2017): 1–11.

29 https://www.ucl.ac.uk/news/2019/apr/key-brain-region-navigating-familiar-places-identified

30 Schumann, Frank, and O'Regan, J. Kevin, 'Sensory Augmentation: Integration of an Auditory Compass Signal Into Human Perception of Space', *Scientific Reports* 7 (2017), https://doi.org/10.1038/srep42197

Chapter 13: The Sex Gap

1 As far as I know, all the studies that I cite in this field are of people who are the gender they were assigned at birth; studies of this type on gender diverse people are sorely lacking.

2 Sorokowski, P., et al., 'Sex Differences in Human Olfaction: A Meta-Analysis', *Frontiers in Psychology*, 10 (2019), https://doi.org/10.3389/fpsyg.2019.00242

3 https://www.perfumerflavorist.com/fragrance/research/The-National-Geographic-Smell-Survey----1The-Beginning-373559131.html

4 Oliveira-Pinto, A. V., et al., 'Sexual Dimorphism in the Human Olfactory Bulb: Females Have More Neurons and Glial Cells Than Males', *PloS ONE*, 9.11 (2014): e111733.

5 https://www.scientificamerican.com/article/fertile-women-heightened-sense-smell/

6 Verma, P., et al., 'Salt Preference Across Different Phases of Menstrual Cycle', *Indian J Physiol Pharmacol*, 49.1 (2005): 99–102.

7 Barbosa, Diane Eloy Chaves, et al., 'Changes in Taste and Food Intake During the Menstrual Cycle', *Journal of Nutrition & Food Sciences*, 5.4 (2015), https://doi.org/10.4172/2155-9600.1000383

8 McNeil, Jessica, et al., 'Greater Overall Olfactory Performance, Explicit Wanting for High Fat Foods and Lipid Intake During the Mid-Luteal Phase of the Menstrual Cycle', *Physiology & Behavior*, 112 (2013): 84–9.

9 Cameron, E. Leslie, 'Pregnancy and Olfaction: A Review', *Frontiers in Psychology*, 5 (2014), https://doi.org/10.3389/fpsyg.2014.00067

10 Choo, Ezen, and Dando, Robin, 'The Impact of Pregnancy on Taste Function', *Chemical Senses*, 42.4 (2017): 279–86.

11 Yoshida, R., 'Hormones and Bioactive Substances That Affect Peripheral Taste Sensitivity', *Journal of Oral Biosciences*, 54.2 (2012): 67–72.

12 Shigemura, Noriatsu, et al., 'Angiotensin II Modulates Salty and Sweet Taste Sensitivities', *Journal of Neuroscience*, 33.15 (2013): 6,267–7.

13 https://www.who.int/whr/2005/chapter3/en/index3.html

14 Kinsley, Craig Howard, et al., 'The Mother as Hunter: Significant *Reduction in Foraging* Costs Through Enhancements of Predation in Maternal Rats', *Hormones and Behavior*, 66.4 (2014): 649–54.

15 Barha, Cindy K., and Galea, Liisa A. M., 'Motherhood Alters the Cellular Response to Estrogens in the Hippocampus Later in Life', *Neurobiology of Aging*, 32.11 (2011): 2,091–5.

16 Keogh, E., and Arendt-Nielsen, L., 'Sex Differences in Pain', *European Journal of Pain*, 8.5 (2004): 395–6, https://doi.org/10.1016/j.ejpain.2004.01.004

17 https://www.nature.com/articles/d41586-019-00895-3

18 McFadden, D., 'Sex Differences in the Auditory System', *Developmental Neuropsychology*, 14.2–3 (1998): 261–98.

19 Fider, N. A., and Komarova, N. L., 'Differences in Color Categorization Manifested by Males and Females: A Quantitative World Color Survey Study', *Palgrave Communications*, 5.1 (2019): 1–10.

20 Abramov, I., et al., 'Sex and Vision II: Color Appearance of Monochromatic Lights', *Biology of Sex Differences*, 31 (2012), https://doi.org/10.1186/2042-6410-3-21

21 https://www.ifst.org/sites/default/files/Is%20Gender%20a%20Challenge%20for%20Your%20Sensory%20Panel%20v9.pdf

Chapter 14: Sensing Emotion

1 James, W., 'What is an Emotion?' *Mind*, 9.34 (1884): 188–205, https://doi.org/10.1093/mind/os-IX.34.188

2 James, W., *The Principles of Psychology*, Henry Holt (1890).

3 Pessoa, Luiz, 'Emotion and Cognition and the Amygdala: From "What is it?" to "What's to be Done?"', *Neuropsychologia*, 48.12 (2010): 3,416–29.

4 Hyman, Steven E., 'How Adversity Gets Under the Skin', *Nature Neuroscience*, 12.3 (2009): 241–3; http://www.columbia.edu/cu/biology/courses/c2006/lectures08/xtra15-08.html

5 Cannon, W. B., 'The James-Lange Theory o Emotions: A Critical Examination and an Alternative Theory' (1927), in the *American Journal of Psychology*, 39: 106–124.

6 Seth, A. K., 'Interoceptive Inference, Emotion, and the Embodied Self', Trends in Cognitive Sciences, 17.11 (2013): 565–73; Seth, Anil K., and Friston, Karl J., 'Active Interoceptive Inference and the Emotional Brain', *Philosophical Transactions of the Royal Society B: Biological Sciences*, 371.1708

(2016), https://doi.org/10.1098/rstb.2016.0007; Seth, Anil K., and Critchley, Hugo D., 'Extending Predictive Processing to the Body: A New View of Emotion?', *Behavioural and Brain Sciences*, 36.3 (2013): 227–8.

7 Nummenmaa, Lauri, et al., 'Maps of Subjective Feelings', *Proceedings of the National Academy of Sciences*, 115.37 (2018): 9,198–203.

8 Critchley, H. D., and Garfinkel, S. N., 'Interoception and Emotion', *Current Opinion in Psychology*, 17 (2017): 7–14.

9 Aristotle, trans. Lawson-Tancred, H. C., *De Anima*, Penguin Classics (1987): 22.

10 Dutton, Donald G., and Aron, Arthur P., 'Some Evidence for Heightened Sexual Attraction Under Conditions of High Anxiety', *Journal of Personality and Social Psychology*, 30.4 (1974): 510–17.

11 Azevedo, Ruben T., et al., 'Cardiac Afferent Activity Modulates the Expression of Racial Stereotypes', *Nature Communications*, 8.1 (2017): 1–9.

12 Nix, Justin, et al, 'A Bird's Eye View of Civilians Killed by Police in 2015: Further Evidence of Implicit Bias', *Criminology & Public Policy*, 16.1 (2017): 309–40; on Black versus white unarmed shootings, see, for example, https://www.washingtonpost.com/investigations/protests-spread-over-police-shootings-police-promised-reforms-every-year-they-still-shoot-nearly-1000-people/2020/06/08/5c204f0c-a67c-11ea-b473-04905b1af82b_story.html

13 Sifneos, P. E, 'Alexithymia, Clinical Issues, Politics and Crime', *Psychotherapy and Psychosomatics*, 69.3 (2000): 113–16.

14 Murphy, J., Catmur, C., and Bird, G., 'Alexithymia is Associated with a Multidomain, Multidimensional Failure of Interoception: Evidence from Novel Tests', *Journal of Experimental Psychology: General*, 147.3 (2018): 398–408.

15 Brewer, R., Cook, R., and Bird, G., 'Alexithymia: A General Deficit of Interoception', *Royal Society Open Science*, 3.10 (2016), https://doi.org/10.1098/rsos.150664

16 Hatfield, E., Cacioppo, J. T., and Rapson, R., *Emotional Contagion*, Cambridge University Press (1994).

17 Dalton, P., et al., 'Chemosignals of Stress Influence Social Judgments', *PLoS ONE*, 8.10 (2013): e77144.

18 Stönner, Christof, et al., 'Proof of Concept Study: Testing Human Volatile Organic Compounds as Tools for Age Classification of Films', *PLoS ONE*, 13.10 (2018), https://doi.org/10.1371/journal.pone.0203044

19 Singh, P. B., et al., 'Smelling Anxiety Chemosignals Impairs Clinical Performance of Dental Students', *Chemical Senses*, 43.6 (2018): 411–17.

20 Gallese, V., et al., 'Action Recognition in the Premotor Cortex', *Brain*, 119.2 (1996): 593–609.

21 For a thorough discussion of mirror neuron research, read Christian Keysers's very engaging *The Empathic Brain*, Social Brain Press (2011).

22 Özkan, D. G., et al., 'Predicting the Fate of Basketball Throws: An EEG study on Expert Action Prediction in Wheelchair Basketball Players', *Experimental Brain Research*, 237.12 (2019): 3,363–73.

23 Dinstein I., et al., 'Brain Areas Selective for Both Observed and Executed Movements', *Journal of Neurophysiolgy*, 98 (2007): 1,415–27; Molenberghs, P., Cunnington, R., and Mattingley, J. B., 'Brain Regions with Mirror Properties: A Meta-Analysis of 125 Human fMRI Studies', *Neuroscience Biobehavioural Reviews*, 36 (2012): 341–9; Mukamel, R., et al., 'Single-Neuron Responses in Humans During Execution and Observation of Actions', *Current Biology*, 20 (2010): R750–6.

24 See Jabbi, M., Bastiaansen, J., and Keysers, C., 'A Common Anterior Insula Representation of Disgust Observation, Experience and Imagination Shows Divergent Functional Connectivity Pathways', *PloS ONE*, 3.8 (2008): e2939.

25 Calder, A. J., et al., 'Impaired Recognition and Experience of Disgust Following Brain Injury', *Nature Reviews Neuroscience*, 3 (2000): 1,077–8.

26 Bird, G., et al., 'Empathic Brain Responses in Insula Are Modulated by Levels of Alexithymia but Not Autism', *Brain*, 133.5 (2010): 1,515–25.

27 Cook, R., et al. 'Alexithymia, Not Autism, Predicts Poor Recognition of Emotional Facial Expressions', *Psychological Science*, 24.5 (2013): 723–32; Bird, G., Press, C., and Richardson, D. C., 'The Role of Alexithymia in Reduced Eye-Fixation in Autism Spectrum Conditions', *Journal of Autism and Developmental Disorders*, 41.11 (2011): 1,556–64.

28 Tottenham, N., et al., 'Elevated Amygdala Response to Faces and Gaze Aversion in Autism Spectrum Disorder', *Social Cognitive and Affective Neuroscience*, 9.1 (2014): 106–117, https://doi.org/10.1093/scan/nst050

29 Garfinkel, Sarah N., et al., 'Discrepancies Between Dimensions of Interoception in Autism: Implications for Emotion and Anxiety', *Biological Psychology*, 114 (2016): 117–26.

30 Spinney, L., 'Consiousness Isn't Just the Brain', *New Scientist* (24 June, 2020).

31 See, for example, Murphy, Jennifer, et al., 'Interoception and Psychopathology: A Developmental Neuroscience Perspective',

Developmental Cognitive Neuroscience, 23 (2017): 45–56; Murphy, J., Viding, E., and Bird, G., 'Does Atypical Interoception Following Physical Change Contribute to Sex Differences in Mental Illness?', *Psychological Review*, 126.5 (2019): 787–9, https://doi.org/10.1037/rev0000158

32 Singer, Tania, and Frith, Chris, 'The Painful Side of Empathy', *Nature Neuroscience*, 8.7 (2005): 845–6.

33 Grice-Jackson, T., et al., 'Common and Distinct Neural Mechanisms Associated with the Conscious Experience of Vicarious Pain', *Cortex*, 94 (2017): 152–63.

34 http://www.alessioavenanti.com/pdf_library/avenanti2006psychneurosci.pdf

35 Maister, Lara, Banissy, Michael J., and Tsakiris, Manos, 'Mirror-Touch Synaesthesia Changes Representations of Self-Identity', *Neuropsychologia*, 51.5 (2013): 802–8.

36 Banissy, Michael J., et al., 'Prevalence, Characteristics and a Neurocognitive Model of Mirror-Touch Synaesthesia', *Experimental Brain Research*, 198.2–3 (2009): 261–72.

37 Banissy, Michael J., and Ward, Jamie, 'Mirror-Touch Synesthesia is Linked with Empathy', *Nature Neuroscience*, 10.7 (2007): 815–16; see also Keysers, C., Kaas, J. H., and Gazzola, V., 'Somatosensation in Social Perception', *Nature Reviews Neuroscience*, 11.6 (2010): 417–28.

Chapter 15: Feeling Sensitive

1 https://hsperson.com/test/highly-sensitive-test/

2 For a round-up of the Arons' early highly sensitive research see Aron, E. N., and Aron, A., 'Sensory-Processing Sensitivity and Its Relation to Introversion and Emotionality', *Journal of Personality and Social Psychology*, 73.2 (1997): 345–68.

3 Wilson, David S., et al., 'Shy-Bold Continuum in Pumpkinseed Sunfish (Lepomis gibbosus): An Ecological Study of a Psychological Trait', *Journal of Comparative Psychology*, 107.3 (1993): 250–60.

4 Aron, Elaine N., Aron, Arthur, and Jagiellowicz, Jadzia, 'Sensory Processing Sensitivity: A Review in the Light of the Evolution of Biological Responsivity', *Personality and Social Psychology Review*, 16.3 (2012): 262–82.

5 Koolhaas, J. M., et al., 'Individual Variation in Coping with Stress: A

Multidimensional Approach of Ultimate and Proximate Mechanisms', *Brain, Behavior and Evolution*, 70.4 (2007): 218–26.

6 Wolf, Max, et al., 'Evolutionary Emergence of Responsive and Unresponsive Personalities', *Proceedings of the National Academy of Sciences*, 105.41 (2008): 15,825–30.

7 Thomas, A., and Chess, S., 'The New York Longitudinal Study: From Infancy to Early Adult Life', in *The Study of Temperament: Changes, Continuities, and Challenges*, Plomin, R. and Dunn, J., eds, Lawrence Erlbaum (1986): 39–52.

8 Kagan, Jerome, 'Temperamental Contributions to Social Behavior', *American Psychologist*, 44.4 (1989): 668–74.

9 Kagan, J., and Snidman, N., *The Long Shadow of Temperament*, Harvard University Press (2009).

10 Boyce, T., *The Orchid and the Dandelion: Why Sensitive People Struggle and How All Can Thrive*, Alfred A. Knopf (2019).

11 Morgan, Barak, et al., 'Serotonin Transporter Gene (SLC6A4) Polymorphism and Susceptibility to a Home-Visiting Maternal-Infant Attachment Intervention Delivered by Community Health Workers in South Africa: Reanalysis of a Randomized Controlled Trial', *PLoS Medicine*, 14.2 (2017), https://doi.org/10.1371/journal.pmed.1002237

12 Kumsta, Robert, et al., '5HTT Genotype Moderates the Influence of Early Institutional Deprivation on Emotional Problems in Adolescence: Evidence from the English and Romanian Adoptee (ERA) Study', *Journal of Child Psychology and Psychiatry*, 51.7 (2010): 755–62; and see Klein Velderman, Mariska, et al., 'Effects of Attachment-Based Interventions on Maternal Sensitivity and Infant Attachment: Differential Susceptibility of Highly Reactive Infants', *Journal of Family Psychology*, 20.2 (2006): 266–74.

13 Pluess, Michael, et al., 'Environmental Sensitivity in Children: Development of the Highly Sensitive Child Scale and Identification of Sensitivity Groups', *Developmental Psychology*, 54.1 (2018): 51–70.

14 Lionetti, Francesca, et al., 'Dandelions, Tulips and Orchids: Evidence for the Existence of Low-Sensitive, Medium-Sensitive and High-Sensitive Individuals', *Translational Psychiatry*, 8.1 (2018): 1–11.

15 Aron, E., The Highly Sensitive Child: Helping Our Children Thrive When the World Overwhelms Them, *Harmony* (2002).

16 Extract reproduced with permission from Carrie Little Hersh's blog, http://www.relevanth.com/when-nature-has-to-conform-to-culture-highly-sensitive-people-in-a-nonsensitive-culture/

17 Chen, X., Wang, L., and DeSouza, A., Temperament, Socioemotional Functioning, and Peer Relationships in Chinese and North American Children', *Peer Relationships in Cultural Context*, Chen, X., French, D. C., and Schneider, B. H., eds (2006): 123–47.

18 Spence, Charles, Youssef, Jozef, and Deroy, Ophelia, 'Where are all the Synaesthetic Chefs?', *Flavour*, 4.1 (2015): 29.

19 http://www.conforg.fr/internoise2000/cdrom/data/articles/000956.pdf

20 https://www.sarahangliss.com/portfolio/infrasonic ; https://www.theguardian.com/science/2003/sep/08/sciencenews.science

21 Persinger, Michael A., 'The Neuropsychiatry of Paranormal Experiences', *Journal of Neuropsychiatry and Clinical Neurosciences*, 13.4 (2001): 515–24.

22 Jung, C. G., 'Psychological Types' (1921), trans. Baynes, H. Godwin, Harcourt, Brace (1923).

23 Aron, Elaine N., *The Highly Sensitive Person*, Thorsons (2017).

24 https://hsperson.com/introversion-extroversion-and-the-highly-sensitive-person/

25 Marco, Elysa J., et al., 'Sensory Processing in Autism: A Review of Neurophysiologic Findings', *Pediatric Research*, 69.8 (2011): 48–54.

26 For research into SPD, see the STAR Institute website: https://www.spdstar.org

27 For work on some shared sensory symptoms between children diagnosed with a disorder and their relatives, see, for example, Glod, Magdalena, et al., 'Sensory Atypicalities in Dyads of Children with Autism Spectrum Disorder (ASD) and Their Parents', *Autism Research*, 10.3 (2017): 531–8.

28 For why people with sensory sensitivities may feel greater empathy, see Acevedo, Bianca P., et al., 'The Highly Sensitive Brain: An fMRI Study of Sensory Processing Sensitivity and Response to Others' Emotions', *Brain and Behavior*, 4.4 (2014): 580–94.

29 For overlap in how SPD and ADHD can present, see, for example, Ghanizadeh, Ahmad, 'Sensory Processing Problems in Children with ADHD: A Systematic Review', *Psychiatry Investigation*, 8.2 (2011): 89–94; https://www.additudemag.com/sensory-processing-disorder-or-adhd/

30 https://www.rescuepost.com/files/library_kanner_1943.pdf

31 Robertson, Caroline E., and Baron-Cohen, Simon, 'Sensory Perception in Autism', *Nature Reviews Neuroscience*, 18.11 (2017): 671–84.

32 See Owen, Julia P., et al., 'Abnormal White Matter Microstructure in Children with Sensory Processing Disorders', Neuroimage: Clinical, 2 (2013): 844–53; Chang, Yi-Shin, et al., 'Autism and Sensory Processing Disorders: Shared White Matter Disruption in Sensory Pathways but

Divergent Connectivity in Social-Emotional Pathways', *PloS ONE*, 9.7 (2014), https://doi.org/10.1371/journal.pone.0103038

33 For more on work linking sensory processing directly to social impairments in autism, see, for example, Thye, Melissa D., et al., 'The Impact of Atypical Sensory Processing on Social Impairments in Autism Spectrum Disorder', *Developmental Cognitive Neuroscience*, 29 (2018): 151–67.

34 Cerliani, Leonardo, et al., 'Increased Functional Connectivity Between Subcortical and Cortical Resting-State Networks in Autism Spectrum Disorder', *JAMA Psychiatry*, 72.8 (2015): 767–7.

35 Puts, Nicolaas AJ, et al. 'Impaired tactile processing in children with autism spectrum disorder.' *Journal of Neurophysiology*, 111.9 (2014): 1803–1811.

36 Kwon, Soo Hyun, et al., 'GABA, Resting-State Connectivity and the Developing Brain', *Neonatology*, 106.2 (2014): 149–55.

37 Parush, S., et al., 'Somatosensory Function in Boys with ADHD and Tactile Defensiveness', *Physiology & Behavior*, 90.4 (2007): 553–8; Puts, Nicolaas A. J., et al., 'Altered Tactile Sensitivity in Children with Attention-Deficit Hyperactivity Disorder', *Journal of Neurophysiology*, 118.5 (2017): 2,568–78.

38 Baron-Cohen, S., et al., 'Prevalence of Autism-Spectrum Conditions: UK School-Based Population Study', *British Journal of Psychiatry*, 194.6 (2009): 500–9; Beau-Lejdstrom, R., et al., 'Latest Trends in ADHD Drug Prescribing Patterns in Children in the UK: Prevalence, Incidence and Persistence', *BMJ Open*, 6.6 (2016): e010508.

39 Green, Shulamite A., and Wood, Emily T., 'The Role of Regulation and Attention in Atypical Sensory Processing', *Cognitive Neuroscience*, 10.3 (2019): 160–2; see also, Ben-Sasson, A., Carter, A. S., and Briggs-Gowan, M. J., 'Sensory Over-Responsivity in Elementary School: Prevalence and Social-Emotional Correlates', *Journal of Abnormal Child Psychology*, 37.5 (2009): 705–16.

Chapter 16: A Sense of Change

1 Barnes, Jonathan, ed., *Complete Works of Aristotle*, vol. 2, Princeton University Press (2014).

2 'Mythbusters: Playing Sound to Plants', https://mythresults.com/episode23; 'Plant Genes Switched on by Sound Waves, *New Scientist* (29 August 2007).

3 Damasio, Antonio, *The Feeling of What Happens*, Vintage (2000); Damasio, Antonio, *Self Comes to Mind*, Vintage (2012).

4 Tanne, Janice Hopkins, 'Humphry Osmond', *BMJ*, 328.7441 (20 March 2004): 713.

5 Huxley, A., *The Doors of Perception*, Harper & Brothers (1954).

6 Tagliazucchi, Enzo, et al., 'Increased Global Functional Connectivity Correlates with LSD-Induced Ego Dissolution', *Current Biology*, 26.8 (2016): R1,043–50.

7 Preller, Katrin H., et al., 'Effective Connectivity Changes in LSD-Induced Altered States of Consciousness in Humans', *Proceedings of the National Academy of Sciences*, 116.7 (2019): 2,743–8. .

8 Roseman, Leor, Nutt, David J., and Carhart-Harris, Robin L., 'Quality of Acute Psychedelic Experience Predicts Therapeutic Efficacy of Psilocybin for Treatment-Resistant Depression', *Frontiers in Pharmacology*, 8 (2018), https://doi.org/10.3389/fphar.2017.00974; Griffiths, Roland R., et al., 'Psilocybin Produces Substantial and Sustained Decreases in Depression and Anxiety in Patients with Life-Threatening Cancer: A Randomized Double-Blind Trial', *Journal of Psychopharmacology*, 30.12 (2016): 1,181–97.

9 The *Mosaic* story on psychedelics as therapeutics is by Sam Wong, https://mosaicscience.com/story/psychedelic-therapy/

Index